Experiments in Analog and Digital Electronics

Text for ECE 3741

Sixth Edition

R. Allen Robinson and Thomas E. Brewer
Georgia Institute of Technology
School of Electrical and Computer Engineering

Kendall Hunt
publishing company

www.kendallhunt.com
Send all inquiries to:
4050 Westmark Drive
Dubuque, IA 52004-1840

ISBN 978-0-7575-9187-7

Printed in the United States of America
10 9 8 7 6 5 4 3 2

Contents

Chapter 1

Introduction and Orientation

1.1 Objective

The objective of this experiment is to provide tyros with an introduction to the instruments which will be used to assemble and evaluate analog and digital electronic circuits. The major items of equipment are the **Tektronix 3012 DPO** 100 MHz oscilloscope, the **Tektronix DMM 4040** Digital Multimeter, and the **CADET** Complete Analog/Digital Electronics Trainer.

1.2 Equipment

1.2.1 Cadet Analog/Digital Electronics Trainer

The acronym **CADET** stands for Complete Analog/Digital Electronics Trainer. This instrument contains DC power supplies, a function generator, logic state indicators, breadboards, switches, etc. that are required to construct analog and digital circuits in a prototype form. It has a power cord which must be plugged into an AC outlet and the on/off switch (upper left hand corner) must be turned on to perform experiments with active devices.

Breadboard

The **CADET** contains three breadboards for circuit construction. These breadboards have rows and columns of holes into which the leads of circuit components such as resistors and integrated circuits can be inserted. Each of these three breadboards has a horizontal trough so that integrated circuits can be inserted with the leads on either side of the trough. The vertical columns on either side of the troughs are electrically common (known as a bus) so that circuit components can be connected to the appropriate lead (called a pin) of an IC. Each of the three breadboards has two horizontal rows above and below the trough area; the holes in each separate row are electrically common except that they are broken in the middle.

DC Power Supplies

Above the three breadboards are four horizontal rows that are connected to the DC voltages shown at the right of these rows. The lowest is the ground bus or zero volts; it is shown connected to a black binding post which is connected to the ground wire in the AC power cord. The next bus is the variable negative (with respect to ground) DC supply which can be adjusted from $-1.3\,\text{V}$ to $-15\,\text{V}$. The next bus is the variable positive DC supply which can be adjusted from $1.3\,\text{V}$ to $15\,\text{V}$. The top bus is the fixed $5\,\text{V}$ supply. The variable supplies are required for linear ICs and the fixed $5\,\text{V}$ supply is required for TTL digital ICs.

Function Generator

On the left hand side of the **CADET** is an output socket which provides continuously variable signals which can be varied in frequency from 0.1 Hz to 100 kHz by changing the setting of a range switch (1, 10 ,100) and a sliding pot (0.1 to 1.0). Four different functions are available: TTL, sine, square, and triangular. TTL is used as the clock input for digital circuits using TTL digital ICs and its amplitude cannot be changed (fixed at 5 V). The amplitude of the other three waveforms can be varied.

Logic Indicators

These are LEDs state indicators located on the right of the **CADET**. They will be used to determine whether the output of a TTL digital circuit is HIGH (+5 V) or LOW (0 V). The switches next to the socket should be set to +5 V and TTL.

Debounced Pushbuttons (Pulsers)

These are located on the left side of the **CADET**. These are debounced switches that are used to step the input to a digital sequential circuit. The acronym PB stands for push-button, the acronym NC stands for normally closed, and the acronym NO stands for normally open. Prior to pushing the button the switch is in the NC position. When it is pushed the switch jumps to the NO position and then back to the NC position.

Potentiometers

The **CADET** contains two variable resistors known as potentiometers or pots for short. They are located at the bottom of the **CADET**. These have a value of 1 kΩ and 10 kΩ. The center two sockets are connected to the wiper of the pot.

BNC Connectors

The **CADET** contains two BNC (Bayonet N Connector) connectors. These will be used to make connections to the oscilloscope. This is a cylindrical lead with the outer lead known as the shield connected to the ground wire in the AC power cord which means that it is connected internally to the ground for the three DC supplies.

SPDT Switches

The **CADET** contains two Single Pole Double Throw switches for general switching applications. Single pole means that one wire is switched and double throw means that the switch has two closed positions.

Logic Switches

Located at the lower left of the **CADET** are eight logic switches. These are used as inputs to the combinational logic experiments. The switch above these eight switches should be set to +5 V.

Speaker

Located on the right of the **CADET** is a speaker. This is a simple speaker which can be used for aural examination of a voltage signal. It has a very poor frequency response.

12.6 VAC

This is located on the upper portion of the **CADET** above the DC power supply strips. This is the output of a center tapped secondary transformer and should not be connected to anything.

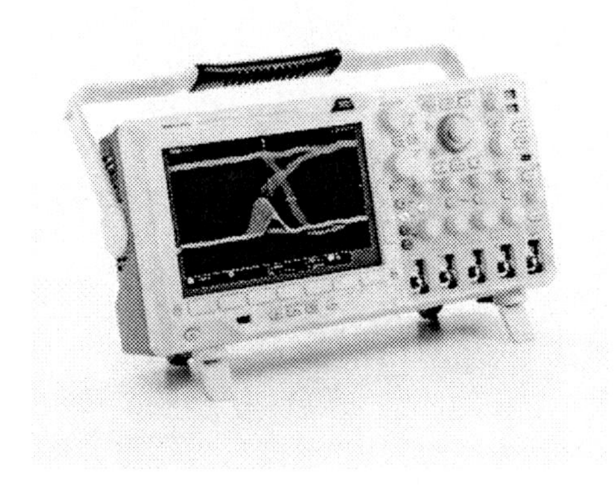

Figure 1.1: Tektronix DPO 3012 Oscilloscope

1.2.2 Tektronix 3012 DPO Oscilloscope

An oscilloscope is an instrument that is primarily used to plot voltage waveforms as functions of time. The **Tektronix 3012 DPO** is a 100 MHz digital oscilloscope. It has two BNC connector inputs for signals and could be used to observe two voltage signals simultaneously. The inputs are digitized, stored in a memory, and then written to a LCD display. The inputs are known as channels and are numbered from **Channel 1** to **Channel 2**. The other front panel BNC connector is an external trigger for the oscilloscope and is used when timing information is required from a signal other that the ones connected to channels 1 and 2. The oscilloscope shown in the above picture has four inputs but other than that the instruments are identical.

This instrument is a digital instrument which means that it contains a microprocessor CPU to enhance measurement capability and decrease the skill required to use it. It contains an auto-scale feature or control which examines the input waveform and automatically sets the vertical and horizontal sensitivity and the trigger controls. Also measurement cursors are available for the measurement of voltage, time, and frequency changes.

The screen of the oscilloscope has a grid overlay. This grid has 10 major horizontal divisions and eight major vertical divisions. The voltage or time that each of the major divisions represents is indicated in an alphanumeric display that appears below and to the left of the LCD display. The time per division is indicated below the display along with the sampling rate and total number of samples.

1.2.3 Tektronix DMM 4040 Digital Multimeter (DMM)

This instrument is a digital multimeter than can be used to measure DC or AC voltage or current or electrical resistance or the frequency of a periodic waveform or signal. The input connectors are known as female banana jack connectors and are located on the upper left; they are named Input HI and LO.

1.3 Procedure

1.3.1 Logic Switches and Indicators

Turn on the **CADET**. The on/off switch is located at the upper left and is on when illuminated. Set the switch above the eight logic switches (lower left) to +5 V. Set the switch above the logic indicators (right side) to +5 V and the switch below to TTL. Use the hook-up wire provided (No. 22 gage single stranded wire) to connect Logic Switch 1 to Logic Indicator 1. If the wire isn't long enough, daisy chain enough wires to make the connection using the buses on the breadboards (never try to insert two or more wires into the same hole). Switch Logic Switch 1 from the 0 to 1 setting and describe what happens to Logic Indicator 1.

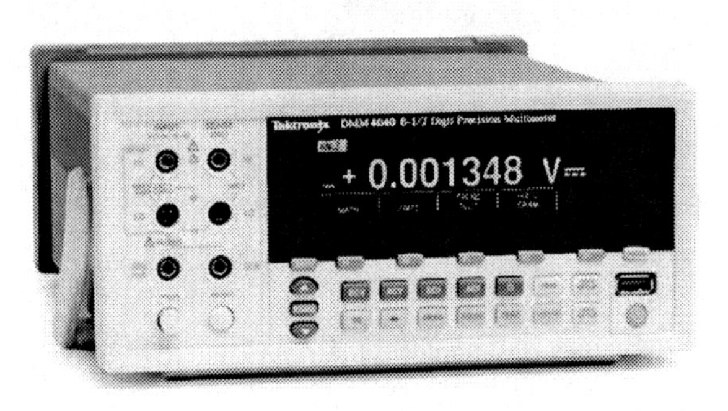

Figure 1.2: Tektronix DMM 4400 Digital Multimeter

1.3.2 TTL Output

Disconnect the connection from Logic Switch 1. Connect the output of the TTL output on the Function Generator (left side and leftmost socket) to Logic Indicator 1. Set the frequency of the function generator to 1 Hz (sliding lever to the upper or 1 setting and two-position switch above to the Hz setting). Describe what happens to Logic Indicator 1 as the frequency of the function generator is varied from 1 Hz to 1 kHz.

1.3.3 Oscilloscope

Connect the TTL output from the Function Generator on the **CADET** to BNC connector 1 (lower left) on the **CADET**. (Disconnect the connection to Logic Indicator 1.) The frequency should be left at 1 kHz. Turn on the Tektronix oscilloscope (push button below and to the left of the LCD display screen with icon of a circle with a line through it) and wait 30 seconds for it to boot.

Connect the BNC connector on the **Channel 1** input on the oscilloscope to the BNC connector connected to the function generator output on the **CADET**. (Do not connect anything to the other inputs of the oscilloscope at this time.) The oscilloscope lead to be used to make this connection is known as a BNC-to-BNC connector; this lead has two male BNC connectors on either end of a coaxial RG58 cable which mates to the female BNC connectors on the oscilloscope and the **CADET**. There is a slot on the male connectors and a ridge on the female connectors and the technique for mating the connectors is to slide the slot on the male connector over the ridge on the female connector and to rotate the male connector 90° clockwise until the connectors snap together.

1.3.4 Autoset

Press **Default Setup** (located below LCD display) **Autoset** on the oscilloscope (button to the far right of the LCD display above the trigger section). A "square wave" should now be displayed on the screen of the oscilloscope. Press the Measure button under the Wave Inspector section of the instrument front panel (located to the right of the LCD screen at the top). Notice that this changes the soft keys just under the CRT display. If any measurements are present, press the tab under Remove Measurements and then Remove All Measurements. Press the soft key for Measure, Add Measurement, rotate the "Multipurpose a" knob (located at the top right of the LCD display) until Peak-to-Peak is highlighted and then OK Add Measurement (located in the lower right of the display). Repeat for Mean, RMS, and Frequency. Press **Menu Off** in the lower right of the display. Record the value of the measurements.

$V_{peak-peak} =$ ______________________________

$V_{mean} =$ ______________________________

$V_{RMS} =$ ______________________________

$Frequency =$ ______________________________

Press **Measure**, **Remove Measurement**, and **Remove All Measurements**. Press **Add Measurement**. Rotate the "**Multipurpose a**" knob until **Snapshot All** is highlighted. Press **OK Snapshot All Measurements**. Note that all of the measurements are displayed but the waveform is obscured. Remove all measurements using the above procedure. Press **Menu Off**.

Press the **Cursor** button to the right of the "**Multipurpose a**" knob. Rotate the "**Multipurpose a**" knob until the first cursor is aligned with the beginning of one period of the waveform. Then rotate "**Multipurpose b**" knob until the second cursor is aligned with the end of a period. Record the difference of the two cursors (the time values) as the period of the waveform.

$Period =$ ______________________________

Press the **Cursor** button until the cursors are turned off. (The button is no longer illuminated.)

Laboratory Instructor Verification ______________________________

1.3.5 RC Circuit

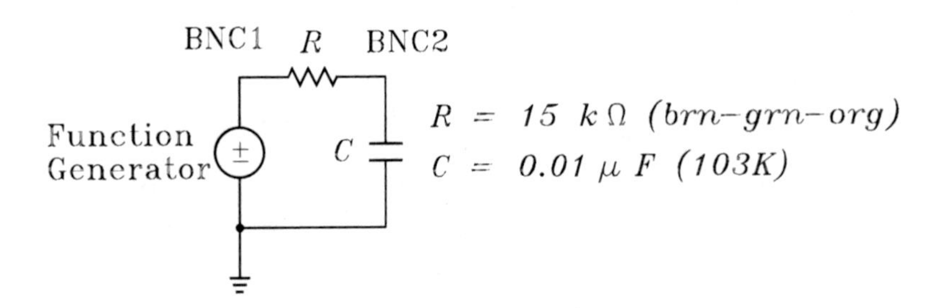

Figure 1.3: RC Circuit

Move the function generator connection to the BNC connector on the **CADET** from TTL to square wave (one hole over). Set the slide switch on the **CADET** to the picture of a square wave. Adjust the amplitude control on the **CADET** until the square wave has a peak-to-peak value of approximately 6 V. Assemble the circuit shown in Fig. 1.3. Note that one side of the capacitor must be connected to the ground bus on the **CADET**. The ground connection shown for the function generator is done internally inside the **CADET** and does not represent a connection that must be made with hook-up wire; simply connect the square output on the function generator to the left side of the resistor. The resistor and capacitor will be supplied by the laboratory instructor. Calculate the frequency

$$f_o = \frac{1}{2\pi RC} \tag{1.1}$$

and set the frequency of the function generator to $f_o/2$. (The easiest way to do this is to turn on the measurements for the frequency and the peak-to-peak voltage for Channel 1.) Vary the frequency controls on the **CADET** until the frequency indicated is approximately $f_o/2$.) Each time the frequency is changed significantly the Autoset button should be pushed.

Note that two oscilloscope channels are being used. The junction between the function generator and resistor is connected to BNC connector No. 1 and the junction between the resistor and capacitor is connected to BNC connector No. 2. Turn Channel 2 on by pressing the **blue** menu button above the BNC connector for Channel 2 .

Press **Autoset**. Press the red **Math** button to the right of the LCD display. Press the **Dual (Wfm) Waveform Math** and toggle through the softkeys to the right of the display until Channel 1 Minus Channel 2 is selected. Turn Channel 1 off by pressing the yellow **menu** button. Vary the vertical position for the two waveforms until they don't overlap. The vertical position for the **Waveform Math** is changed by pressing the **Waveform Math** button and then "**Multipurpose a**" knob controls the position and "**Multipurpose b**" controls the volts/division. The vertical position of Channel 2 is varied by the vertical position button for Channel 2.

1.3.6 Printing the Oscilloscope Display with a PC

Insert a USB memory stick into the USB slot on the front of the oscilloscope. Press the Save button to the right of USB slot. The display is now stored on the memory stick and can be printed from any pc.

1.3.7 Theoretical Predictions

The resistor voltage is the discontinuous waveform and the capacitor voltage is the continuous waveform. When the square wave is positive the theoretical expression for the resistor voltage is

$$v_R(t) = 2A\ e^{-t/RC} \tag{1.2}$$

and the capacitor voltage

$$v_c(t) = 2A[1 - e^{-t/RC}] - A \tag{1.3}$$

where A is the peak value of the square wave. How do well do these waveforms compare with these equations?

____

____

Display the properly functioning circuit to the laboratory instructor.

Laboratory Instructor Verification________________________

1.3.8 Sine Wave Response

Turn off the **Math function** by pressing the red **Math** button. Turn Channel 1 on by pressing the **yellow** menu button above the BNC connector for Channel 1. Switch the function on the function generator on the **CADET** to sine. Press **Autoset**. Press **Measurement**, **Add Measurement**, the **Peak-to-Peak** voltage for Channel 1 and Channel 2. (To obtain the **Peak-to-Peak** value for Channel 2, rotate the "**Multipurpose b**" knob until Channel 2 is selected. Vary the frequency of the function generator from 0.1 f_o to 10 f_o and measure and record the peak to peak value of the input V_1 and output V_2 sinusoidals as well as the frequency on the data sheet. Each time the frequency is changed press the **Autoset** control.

Frequency	(Hertz)	V_1 (Volts)	V_2 (Volts)
$0.1f_o$			
$0.3f_o$			
$0.7f_o$			
f_o			
$3f_o$			
$7f_o$			
$10f_o$			

Display the properly functioning circuit to the laboratory instructor.

Laboratory Instructor Verification________________________

Plot this V_2/V_1 versus f on the attached sheet of log-log graph paper. (Alternatively use the pc to plot the graph using a spreadsheet such as Excel or K Graph. Change both the x and y axes of the plot from linear to log.) This circuit is known as a low-pass filter for which the theoretical value of this voltage ratio should be

$$\frac{V_2}{V_1} = \frac{1}{\sqrt{1 + (2\pi fRC)^2}} \tag{1.4}$$

How well does this plot agree with this equation?

1.4 Laboratory Report

Turn off both the **CADET** and the oscilloscope and return the resistor, capacitor, and hook-up wire to the laboratory instructor. Turn in the answers to all questions and the requested sketches and plots. Answer any supplementary questions that may be posed by the laboratory instructor.

Title:______________________________

Figure No.____________________________

VOLTS/DIV

Ch1 __________
Ch2 __________
Ch3 __________
Ch4 __________

TIME/DIV
MAIN

____

DELAYED

____

DELAY

____

Coupling

AC ☐
DC ☐
Ch 2 Inv ☐
BW Limit ☐

Title:______________________________

Figure No.____________________________

VOLTS/DIV

Ch1 __________
Ch2 __________
Ch3 __________
Ch4 __________

TIME/DIV
MAIN

____

DELAYED

____

DELAY

____

Coupling

AC ☐
DC ☐
Ch 2 Inv ☐
BW Limit ☐

Chapter 2

Common Emitter Circuit

2.1 Objective

The objective of this experiment is to examine some elementary common emitter circuits. A single NPN BJT is used as the active device. The circuit is used as an amplifier and as a switch.

2.2 Theory

Transistors are three terminal circuit elements where the current or voltage at one terminal is used to control the current or voltage at another terminal. There are two basic types of transistors, bipolar and field effect with field effect being further subdivided into junction and insulated gate transistors. Transistors are fundamental components in all modern analog and digital electronic circuits and systems. The analog and digital integrated circuits used in subsequent experiments in this course contain numerous transistors within the packages. This experiment examines the use of a bipolar transistor as an active element in an electronic amplifier and a voltage controlled switch.

2.2.1 NPN BJT Transistor

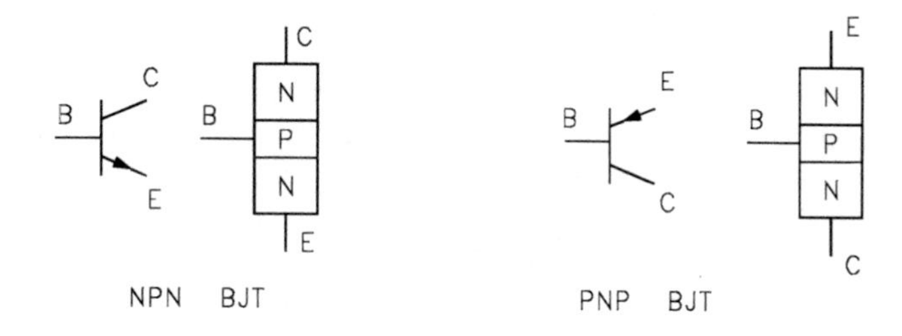

Figure 2.1: Transistor symbols.

The circuit symbol for an NPN BJT (bipolar junction transistor) is shown in the far left in Fig. 2.1. There are three terminals known as the emitter, the base, and the collector. When used as an amplifier dc current flows into the collector and base terminals and out of the emitter with the arrow on the emitter terminal indicating the direction of current flow. Normally the base current is much smaller than the current flowing in the other two terminals. Small changes in the base current can cause much larger changes in the current in the other two terminals which makes it possible to use it in either an amplifier or a switch.

Transistors are solid state devices which are fabricated out of semiconductor silicon crystals. Impurities known as dopants are added to a wafer of intrinsic or pure silicon that produce only minute changes in the chemical composition of the structure but greatly alter its electrical characteristics. A conceptual representation of this is shown to the right of the circuit symbol. Two types of impurities known as p and n are used to increase the number of free holes or free electrons in the crystalline structure and the boundary between two such regions is known as a pn junction. There are three regions in the crystalline structure known as the emitter (n type impurity) , base (p type impurity), and collector (n type impurity) with the base sandwiched in between the emitter and collector. The physical width of the base region is much smaller than either the collector or emitter. Although the types of doping added to the collector and emitter are identical (n type) the doping levels and physical dimensions are considerably different so that these two terminals are not interchangeable.

The complement of the NPN BJT is the PNP BJT shown on the right in Fig. 2.1 along with the conceptual depiction of the arrangement of the three areas in the silicon crystalline structure. The arrow on the emitter again indicates the direction of flow of dc current through the device. For the NPN device current flows into the base and collector terminals and out of the emitter whereas the direction of each current is reversed for the PNP device. Oftentimes both types of transistors are needed in an electronic circuit.

When the transistor is used as an amplifier the collector-base pn junction is reverse biased and the emitter-base junction is forward biased. This means that the voltages V_{BE} and V_{CB} are positive for the NPN device. When this is the case a small change in the base current is greatly magnified or amplified as a change in the collector current. This provides current gain as described by

$$\beta = \frac{i_C}{i_B} \tag{2.1}$$

where i_B is the current flowing into the base terminal and i_C is the current flowing into the collector. The parameter β is dimensionless and is usually large. For the type of transistor to be used in this experiment it has a value between one and two hundred. This means that the collector current and emitter currents are essentially the same; these are usually in the milliampere range whereas the base current is in the microampere range. The collector current is related to the emitter current by

$$\alpha = \frac{i_C}{i_E} \tag{2.2}$$

which is a factor slightly less than unity.

This type of transistor is known as a bipolar transistor because the current flowing through the device is conducted by two different types of charge carriers, free electrons and free holes. This is in contrast to the field effect transistors which are know as unipolar since they use only one type of charge carrier. Field effect transistors such as insulated gate MOS (metal oxide semiconductor) or MOSFET are often preferred in digital logic because they dissipate less power and therefore produce less heat which means that more devices can be packed into a semiconductor chip. Two types of MOS transistors, know as P Channel and N Channel, are often paired together to form CMOS devices which are used exclusively in certain logic families and computer memories. FETs are used where power dissipation is the dominant concern but when power gain or switching speed is the criteria BJTs have no peer.

The terminal characteristics of the NPN BJT depend on the voltages applied to the terminals. When the voltage from the base to the emitter is negative or positive and less than a voltage known as the on-voltage ($V_{BE(on)}$) the current is essentially zero. The on-voltage for the transistor has a value of approximately 0.65 V for the transistor that will be used in this experiment. Once the voltage from base to emitter exceeds the on-voltage and the collector to base junction is reversed bias by 1 V or more, the current flowing into the collector increases exponentially with the base to emitter voltage until limited by the external circuit.

2.2.2 Common Emitter Circuit

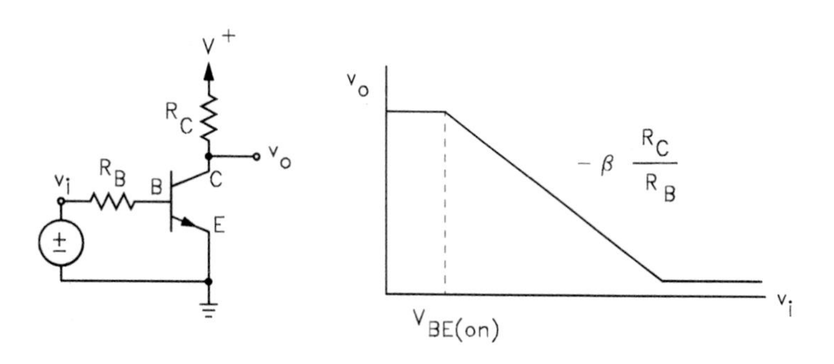

Figure 2.2: Common emitter circuit.

The circuit shown in Fig. 2.2 is known as a common emitter circuit because the emitter terminal of the transistor is common to the input voltage, v_i, and the output voltage, v_o. The ground symbol in the diagram (three horizontal lines of decreasing width) simply indicates a reference point for zero volts in the circuit and does not necessarily indicate a connection to the earth ground wire in the AC outlet. All voltage which are indicated with only one subscript are understood to be referenced to ground. The symbol with the up arrow that has V^+ next to it means that the positive terminal of the dc power supply is to be connected to it. This arrangement means that a dc power supply is connected between the up arrow and ground; to simplify the diagram the dc power supply isn't shown. This dc voltage reverse biases the collector to base pn junction. The lead with v_o next to it is just a wire with nothing attached to it.

A plot of v_o versus v_i is to the right of the circuit diagram in Fig. 2.2. For values of v_i from 0 to $V_{BE(on)}$ the base to emitter pn junction is off and no current flows from the power supply to ground. This makes the voltage drop across the resistor R_C zero which makes the output voltage V^+. Once the input voltage equals or exceeds $V_{BE(on)}$ current begins to flow and the output voltage is given by

$$V_o = V^+ - I_C R_C = V^+ - \beta I_B R_C = V^+ - \beta \left(V_i - V_{BE(on)}\right) \frac{R_C}{R_B} \tag{2.3}$$

which is the portion of the plot which is a straight line with a slope of $-\beta R_C/R_B$. Once the voltage drop across R_C reaches V^+ the output voltage is zero and cannot decrease any further. Further increases in the base current do not increase the collector current and the transistor is said to be saturated. The collector to emitter voltage doesn't drop to exactly zero volts but a voltage of a few tenths of a volt known as the collector to emitter saturation voltage.

If the circuit is to be used as an amplifier the output voltage must be restricted to that portion where is the slope is $-\beta R_C/R_B$. To use the circuit as a switch or logic inverter the output is confined to either the cutoff or saturation regions.

2.2.3 Common Emitter Amplifier

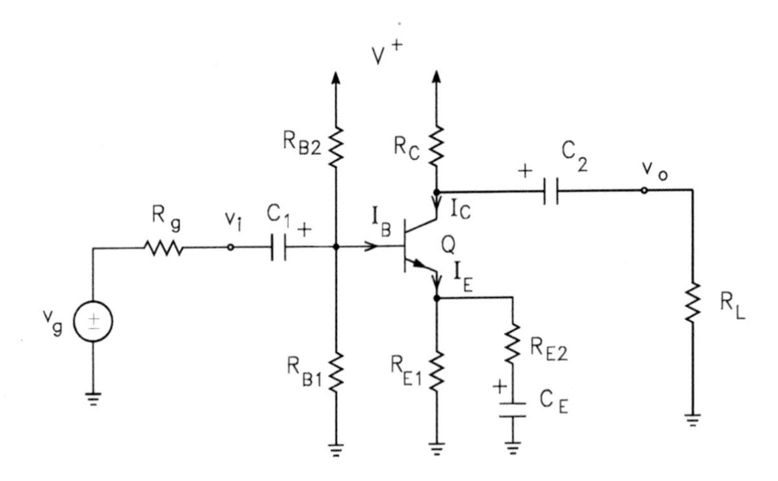

Figure 2.3: Common emitter amplifier.

The circuit shown in Fig. 2.3 is known as a single stage common emitter amplifier. Since there is only one active device there is only one stage of voltage gain. The input is the voltage source, v_i, and the output is the voltage that appears across the load resistor, v_o. The goal of the amplifier is to make the output an undistorted larger version of the input signal. The input is supplied by a function generator with a Thévenin voltage v_g and a source resistance R_g. It will be assumed that R_g is small so that $v_i \approx v_g$.

The resistors R_{B1} and R_{B2} bias the circuit so that it is in the linear region so that output will not lie in the cutoff or saturation portion of the characteristic for sufficient small values of the input. The capacitors C_1 and C_2 are dc blocking capacitors that prevent dc from appearing on the input, v_i, or the output, v_o, i.e. the dc current flowing in the load resistor and function generator is zero. The capacitor C_E is an emitter bypass capacitor that increases the gain of the amplifier at signal frequencies (frequencies for which the capacitors are short circuits). Each of the capacitors shown in the circuit diagram are assumed to be open circuits at dc and short circuits at signal frequencies. The dc value of the collector current is given by

$$I_C = \frac{V^+ \dfrac{R_{B1}}{R_{B1}+R_{B2}} - V_{BE(on)}}{R_{E1} + (R_{E1} + R_{B1}\|R_{B2})/\beta} \tag{2.4}$$

where the resistors are selected to bias the circuit to achieve a desired quiescent collector current (the collector current when $v_i = 0$.)

It is assumed that the input v_i is a sine wave with a frequency such that the capacitors C_1, C_2, and C_E can be assumed to be short circuits. The input is expressed as a rms phasor V_i and the output as a phasor V_o. The voltage gain of the circuit is given by

$$A_v = \frac{V_o}{V_i} = -\alpha \frac{R_C\|R_L}{r_e + R_{E1}\|R_{E2}} \tag{2.5}$$

where $r_e = V_T/I_E$ where V_T is the thermal voltage which is 25.9 mV when $T = 300\,\mathrm{K}$ and $\alpha = \beta/(\beta+1)$. The resistor R_{E2} is used to adjust the gain without affecting the bias. The minus sign in Eqn. 2.5 simply means that the output sine wave is 180° out of phase with the input sine wave which makes this an inverting amplifier. Eqn. 2.5 is correct for signal frequencies; the gain will decrease at low frequencies as the coupling and bypass capacitors impedance will no longer be negligible and will also decrease at high frequencies due to junction capacitances inside the transistor. However, for a large band of frequencies, known as the midband, the gain will be a constant and given by 2.5.

As the amplitude of the input is increased eventually the amplifier will leave the linear region and the output will be distorted on either the positive or negative peak of the waveform. If the input is made very large the top and bottom of the waveform will be appear to be a flat line. This a form of severe distortion known as clipping. It corresponds to the transistor leaving the linear region and entering either the cutoff or saturation mode.

Normally the load resistor, R_L, and gain, A_v, are specified and the other parameters can be selected by the circuit designer. A desirable feature of any amplifier circuit is that it produce the maximum undistorted swing in the input signal that can be achieved for a particular dc power supply voltage. This is done by biasing the circuit for symmetric clipping. Symmetric clipping means that the output begins to clip on the positive and negative peak for the same value of the input.

The distortion produced by an amplifier can be determined with a spectrum analyzer. This is an instrument that displays the frequency spectra of a signal. If a sine wave is connected to the input of a distortionless amplifier then the output should be a sine wave with a larger amplitude but there should not be spectral components at other frequencies. When the input becomes so large that the amplifier distorts the signal, significant spectral components can be seen at other frequencies which are integral multiplies of the input frequency.

2.2.4 Common Emitter Switch

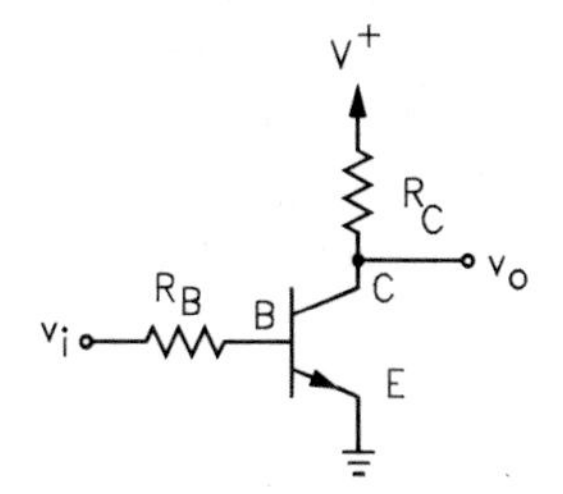

Figure 2.4: Common emitter switch.

A circuit diagram of a common emitter switch is shown in Fig. 2.4. It is assumed that the input voltage signal is that used in TTL digital logic, i.e. 0 V known as a LOW which could represent a logical 0 and 5 V known as a HIGH which could represent a logical 1. The output voltage is the voltage at the collector of the transistor. The power supply voltage is $V^+ = 5\,\text{V}$.

When the input is LOW (0 V) the base to emitter junction is not on and the current flowing through the transistor is zero. This makes the voltage drop across R_C zero which makes the output $V^+ = 5\,\text{V}$ or HIGH. When the input is HIGH the transistor is driven into saturation which reduces the collector to emitter voltage to essentially zero or LOW. (The resistor R_B is picked to be small enough so that an input of 5 V produces enough base current to saturate the transistor.) Thus this circuit performs a logical inversion. It is therefore an INVERTER or NOT gate. Elementary variations of this circuit can also produce NAND or NOR gates.

The collector to emitter terminal is either an open circuit or a short circuit depending on the input. This makes this circuit a voltage controlled switch. The switch is either open or closed depending on the input voltage.

Actual logic inverters use more than one transistor to improve the switching speed and reduce the power dissipation. But the principle of switching a transistor from saturation to cutoff is still use to implement the logic function.

A visible indication of the state of the inverter can be obtained by placing a light emitting diode (LED) in series with the collector. The diode is oriented so that when the collector to emitter voltage is LOW current flows through the diode which causes it to emit light.

2.2.5 Diode

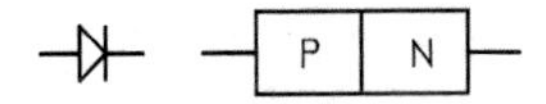

Figure 2.5: Diode.

A diode is a two terminal circuit element. It is a solid state device fabricated from a wafer of silicon. Dopants are added to form p and n type regions. It has only one pn junction whereas transistors have two. It permits current to flow with little or no resistance from p to n but presents a very high resistance to the flow of current from n to p.

The circuit symbol is shown in Fig. 2.5. The p side is the triangle and the n side is the straight line. It would be inserted in Fig. 2.4 in series with the resistor R_C with the p side on top.

When current flows through a pn junction holes and electrons collide and emit photons of light. If the semiconductor is silicon the light isn't visible to humanoids. For an LED the semiconductor used is GaAs. Depending on the type of impurity added the LED can produce red, green, yellow, or amber light. When current is flowing from p to n the voltage drop across a silicon pn junction is about 0.7 V but for GaAs it is in the range of 1.5 V.

2.3 Procedure

2.3.1 Junction Resistance Measurement

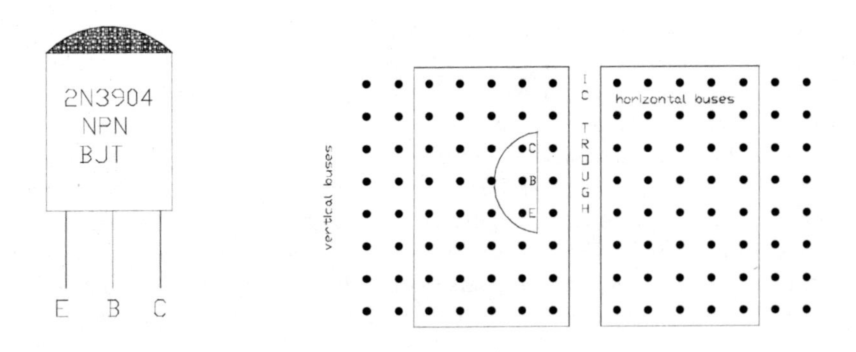

Figure 2.6: Lead arrangements for 2N3904 NPN BJT.

The active device used in this experiment is the 2N3904 NPN BJT. The lead arrangement for this three terminal device is shown in Fig. 2.6. When held so that the leads point down and the flat face faces the experimenter the leads are, from left to right, emitter (E), base (B), collector (C).

Insert the transistor onto the breadboard as shown in Fig. 2.6. Choose the center breadboard section on the **CADET**. The three leads are inserted so that each is in a separate adjacent parallel horizontal bus. Place the emitter at the bottom. Insert a wire into a hole on each of the three buses so that voltages can be measured between the terminals of the transistor.

Turn on the **Tektronix DMM 4040** Digital Multimeter and wait for it to boot. Press the symbol for the diode on the second row (triangle touching a vertical line). This configures the instrument to measure the voltage or "resistance" of a pn junction. Leads should be inserted into the HI and LO jacks on the multimeter in the upper left where the connectors are labeled INPUT.

Connect the HI lead to the base wire and the LO lead to the emitter wire and record the meter reading

V_{BE} =________________________

Connect the HI lead to the emitter and the LO lead to the base and record the meter reading

V_{EB} =________________________

Connect the HI lead to the base and the LO lead to the collector and record the meter reading

V_{BC} =________________________

Connect the HI lead to the collector and the LO lead to the base and record the meter reading

V_{CB} =________________________

When set to the diode checker position the meter will indicate a voltage of around 0.7 V when the junction is forward biased by the multimeter and an open circuit when reversed biased. Although what is actually being measured is a voltage, the term that is normally used is junction "resistance". If the readings made above do not indicate a high resistance with one orientation and a small resistance with the other, the transistor is probably defective.

Laboratory Instructor Verification ____

2.3.2 Bias

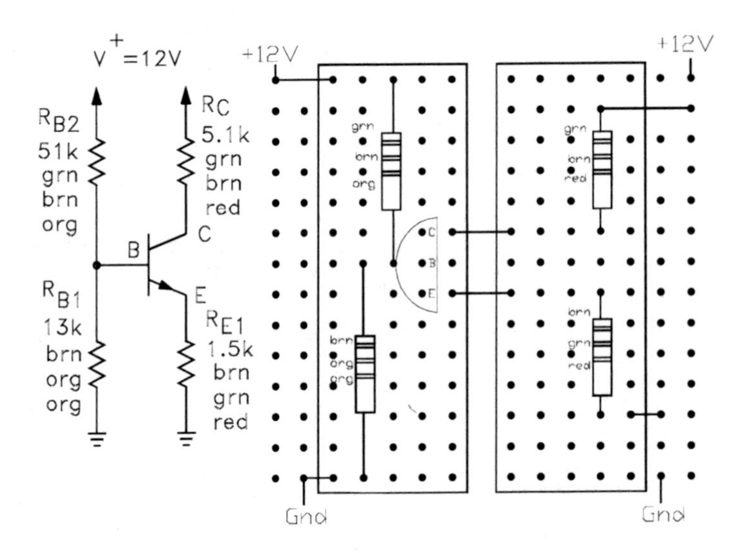

Figure 2.7: Bias circuit.

Assemble the circuit shown in Fig. 2.7 with the power off on the **CADET**. Both a standard electronic circuit diagram as well as a depiction of one implementation of this circuit are shown. Begin by laying out buses or rails for +12 V (which will come from the $+V$ variable dc supply at the top of the **CADET**) and ground (also found at the top of the **CADET**). Since the vertical buses are broken halfway down, four vertical jumpers should be used to extend the vertical buses from the top to the bottom of the breadboard. For ease of implementation two + 12 V and ground buses will be used. As shown the two outside vertical rails are +12 V and the two inner are ground.

Insert the resistors as shown. If there is a question about a resistance value press the button on the DMM to implement a 2 wire resistance measurement and measure the resistance value (Ω 2W Input). Be careful not to touch both leads of the ohmmeter with the experimenter's digits since this will cause the resistance measured to be the parallel combination of the resistor and the experimenter.

Turn on the **CADET**. Press the button DCV on the **Tektronix DMM 4040** DMM. Connect the HI lead to the $+V$ bus and the LO lead to the ground bus. Measure the dc voltage from the $+V$ bus to ground and adjust it using the $+V$ control on the **CADET** until it is approximately +12 V. Record the value measured by the DMM.

$V^+ =$ ____________________

With the LO lead connected to the ground rail, measure the voltage at the collector, base, and emitter terminals by successively touching the HI lead to each terminal.

$V_C =$ ____________________

$V_B =$ ________________________

$V_E =$ ________________________

Calculate the value of the collector current as

$$I_C = \frac{V^+ - V_C}{R_C} = \underline{\hspace{5cm}} \tag{2.6}$$

Compare this with the theoretical value predicted by Eqn. 2.4. For this calculation assume that $\beta = 100$ and $V_{BE(on)} = 0.65\,\text{V}$. If these two are not reasonably close, there is a problem with the circuit which must be corrected prior to going to the next step in the experiment.

Laboratory Instructor Verification ________________________

2.3.3 Common Emitter Amplifier

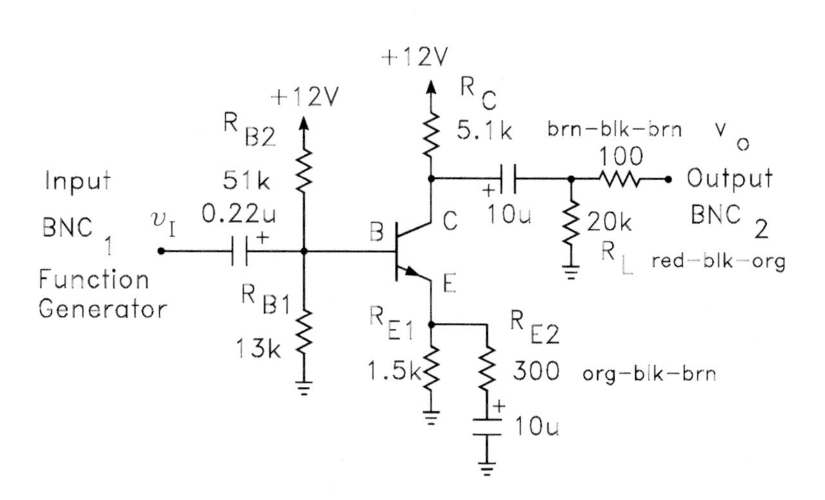

Figure 2.8: Common emitter amplifier.

Assemble the circuit shown in Fig. 2.8 with the power on the **CADET** turned off. This is simply the previous circuit with some additional components added. The left terminal is the input to the circuit which comes from the function generator on the far left of the **CADET;** be sure to use the breadboard sockets for the function generator sine, triangular, or square output and not the TTL output. The output of the function generator is also connected to BNC connector 1 in the lower left of the **CADET**. The output of the circuit is to be connected to BNC connector 2 in the lower right of the **CADET**. The $100\,\Omega$ resistor at the output of the circuit does not appear in the theory section and is inserted for practical reasons; namely, the oscilloscope lead has a capacitive input which might cause the circuit to oscillate (a signal unrelated to the input) and this resistor reduces the possibility of that occurring.

Three of the components are capacitors which are used to block dc. The capacitor C_1 is a ceramic disk capacitor which will say $224k$ or $0.22\,\mu$F and has no polarity associated with it, i.e. it may be inserted into the circuit with either polarity. A polarity marking is shown for completeness. The other two capacitors are electrolytic capacitors which have a polarity associated with them. These are small metal cans with two leads coming out of them and one will be indicated as the negative lead. The capacitance value will be printed on them. They must be inserted into the circuit with the proper polarity or they will not function as capacitors. Indeed, if they are inserted with the incorrect polarity they may

EXPLODE!!!!!!!

causing irreversible and catastrophic damage to the capacitors and angst to the experimenter.

Connect BNC connectors 1 & 2 on the **CADET** to Ch1 and Ch2 on the **Tektronix 3012 DPO** oscilloscope. Turn on the oscilloscope and wait for it to boot. Turn on the **CADET** and adjust the function generator to produce a 1 kHz sine wave. On the oscilloscope, press **Default Setup**, turn Ch 2 on by pressing the blue **Menu** button and press **Autoset** on the oscilloscope. Adjust the amplitude of the sine wave produced by the function generator until the output of the circuit (waveform on $CH2$ of the oscilloscope) is a sine wave with a peak-to-peak value of approximately 2 V. It may be necessary to press **Autoset** as the level and frequency are changed.

Measure the peak to peak value of the sine waves on $CH1$ & $CH2$. Press **Measure**, **Add Measurement**, rotate the "**Multipurpose a**" knob until frequency is highlighted, **OK Add Measurement** on the oscilloscope. Add the measurements for the peak-to-peak value of the voltages on Ch 1 & Ch2. (To switch the source for measurments from Ch1 to Ch2 rotate the "**Multipurpose b**" knob.) Print the display by inserting a USB memory stick into the port on the front and pressing the **Save** button to the right of the stick.

$V_1 = V_i = V_g =$ ____________________

$V_2 = V_o =$ ____________________

Calculate the gain as

$$|A_v| = \frac{V_2}{V_1} = \_\_\_\_\_\_\_\_ \tag{2.7}$$

Compare this with the theoretical result given by Eqn. 2.5; assume that $I_E = I_C$ for this equation and use the value of the collector current that was measured. The minus sign in Eqn. 2.5 merely indicates that the input and output waveforms are 180° out of phase.

Laboratory Instructor Verification ____________________

2.3.4 Large Signal Behavior

Increase the amplitude of the sine wave produced by the function generator until the output becomes distorted (no longer looks like a sine wave with the top and bottom cut off or clipped). Turn Ch 1 off by pressing the yellow Menu button. Turn the cursors on (toggle the button to the right of the "**Multipurpose a**" knob. Use "**Multipurpose a**" knob to position the first cursor at the positive peak. Use "**Multipurpose b**" knob to position the second cursor at the negative peak.

Print the display. Turn the cursors off.

Laboratory Instructor Verification ____________________

2.3.5 Spectral Analysis

Adjust the amplitude of the sine wave produced by the function generator until the amplitude of the output of the circuit is a undistorted sine wave with a frequency of 1 kHz. Press Autoset. Turn Ch1 off if it is on.

The oscilloscope will now be configured to display the frequency spectrum of the waveform connected to the Ch2 input which is the output of the circuit. Vary the HORIZONTAL SCALE until the TIME/DIV is set to 2 ms which will display 20 cycles of the waveform on the screen. Use the VERTICAL POSITION and SCALE until the waveform is position in the upper half of the screen and occupies approximately 1 major division. Do not position the waveform off the screen; this will clip it and produce a severely distorted spectrum. Press the red **Math** button followed by **FFT**. Set the source for the FFT to CH2 (Change the source for the FFT with the "**Multipurpose a**" knob.) Use the VERTICAL SCALE and POSITION controls for the FFT until the spectra is position in the lower half of the display and occupies about 3 major vertical divisions. (Press **Math**, **FFT**, then "**Multipurpose a**" button controls the position and "**Multipurpose b**" controls the scale.) Turn Ch2 off by pressing the blue **Menu** button. Rotate the Horizontal Scale knob until the frequency per division for the spectra is 1.25 kHz.(Indicated in red.) A spike approximately one horizontal division to the right of the left most graticule line should be visible. Print the display.

Increase the amplitude of the input signal until the output distorts or clips. There should be a spike at $f = 1\,\text{kHz}$ which is the amplified input 1 kHz sine wave. This is the fundamental. The other spectral components are distortion components. Vary the amplitude of the sine wave produced by the function generator and observe the spectrum of the output. Adjust the amplitude so that all distortion components are just below the digital noise baseline. Print display. Turn the *Math* function off and Ch1 and Ch2 on.

Press Auto-scale and measure the amplitudes of the waveforms. Print the display.

Laboratory Instructor Verification ____________________________

2.3.6 Common Emitter Switch or INVERTER

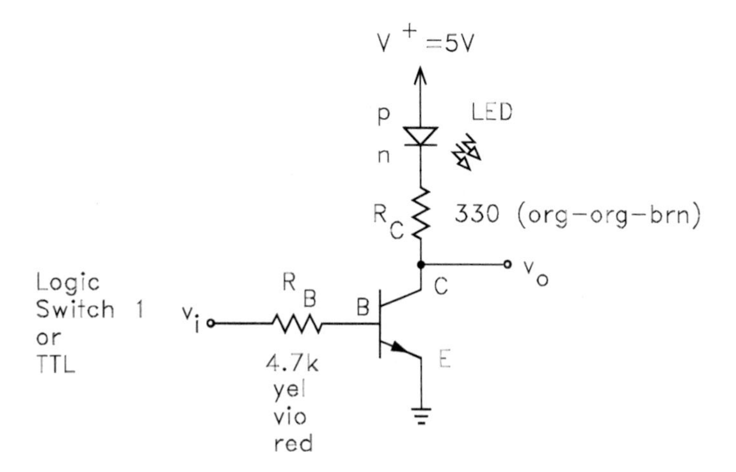

Figure 2.9: INVERTER.

Assemble the circuit shown in Fig. 2.9 with the power on the **CADET** turned off. The input v_i is to come from logic switch 1 (LS1) at the bottom of the **CADET**. Don't connect anything to v_o.

The diode shown (triangle touching a vertical line) is a light emitting diode (LED). The polarity of the leads must be determined. Turn on the **Tektronix DM4040** DMM and wait for it to boot. Press the button with the symbol for a diode. Connect the HI lead to one LED lead and the LO to another. When the HI is connected to the p side (triangle) and the LO to the n side (line) the diode will light. The opposite connection of the leads will produce a dark diode.

Turn on the **CADET** and observe the effect of switching logic switch one from the high to low position. (The switch next to logic switch 1 should be set to +5 and not to $+V$.)

Turn the power off and connect v_i to the TTL output of the function generator (remove the connection to logic switch 1) and also connect the TTL output to BNC1 and the output v_o to BNC2. Connect BNC1 & BNC2 to Chs 1 & 2 of the oscilloscope. Turn the **CADET** on and set the frequency of the TTL output to 1 kHz. Press **Autoset**. Print the display.

Reduce the frequency of the TTL signal to 10 Hz and then 1 Hz and observe the LED.

Laboratory Instructor Verification ______________________________

2.4 Laboratory Report

Turn is all plots, calculations, and answer any questions posed in the procedure as well as any supplemental questions by the laboratory instructor.

Turn everything off, return the BNC to BNC leads to the wall rack, and return any components supplied by the laboratory instructor to the laboratory instructor.

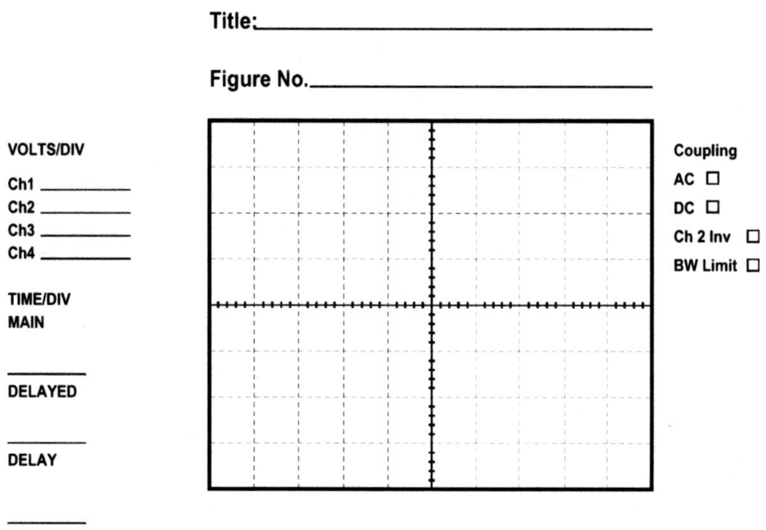

Title:____________________

Figure No.____________________

VOLTS/DIV

Ch1 ________
Ch2 ________
Ch3 ________
Ch4 ________

TIME/DIV
MAIN

____
DELAYED

____
DELAY

____

Coupling
AC ☐
DC ☐
Ch 2 Inv ☐
BW Limit ☐

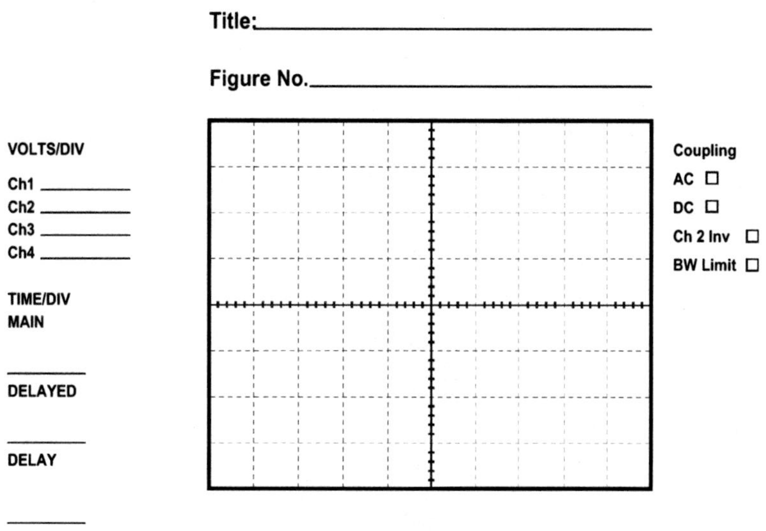

Title:____________________

Figure No.____________________

VOLTS/DIV

Ch1 ________
Ch2 ________
Ch3 ________
Ch4 ________

TIME/DIV
MAIN

____
DELAYED

____
DELAY

____

Coupling
AC ☐
DC ☐
Ch 2 Inv ☐
BW Limit ☐

Figure 2.10:

Chapter 3

Op-Amp Amplifiers

3.1 Objective

The objective of this experiment is to experimentally examine one of the fundamental building blocks of analog electronics, the op-amp amplifier. The amplifier configurations that will be examined are the noninverting, inverting, differential, and instrumentation op-amp amplifier.

3.2 Theory

The measurement of physical parameters such as pressure, temperature, displacement, fluid flow, etc. usually begins with a transducer which is a device that produces an electrical voltage that is a function of the physical variable being measured. Electrical signals are preferred because they are easy to transmit, amplify, store, and numerically process with a computer. Because the amplitude of the voltages produced by a transducer are usually low, it is necessary to amplify them to a level that can be processed by instrumentation. The op-amp amplifier is an integrated circuit amplifier that is commonly used for this purpose.

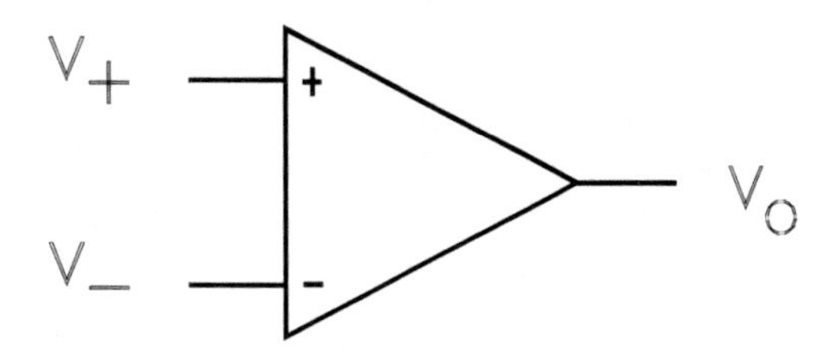

Figure 3.1: Ideal op-amp.

Shown in Fig. 3.1 is the circuit symbol for an ideal op-amp. It has two inputs and one output; the inputs are on the left and the output is on the right. The symbol is a triangle with a plus sign next to the input which is known as the noninverting input and a minus sign next to the input known as the inverting input. The output is on the right side of the triangle and is given as a function of the inputs by

$$v_o = A_{OL}(v_+ - v_-) \tag{3.1}$$

where A_{OL} is known as the open loop gain. The ideal op-amp has an infinite input impedance (no current flows into either the inverting or noninverting input), an infinite gain ($A_{OL} \to \infty$), an infinite bandwidth (A_{OL} is not a function of frequency), zero DC bias currents (a problem with physical op-amps which causes small DC currents to flow into the inputs), zero DC offset voltage (a problem with physical op-amps which causes a small DC voltage to appear at the output), and an infinite slew rate (v_o responds instantaneously to a change in either input). Op-amps are never, well hardly ever, used without negative feedback which makes $v_+ = v_-$ if v_o is finite and A_{OL} is infinite. For the experiments that will be performed in this course, the ideal op-amp model is adequate.

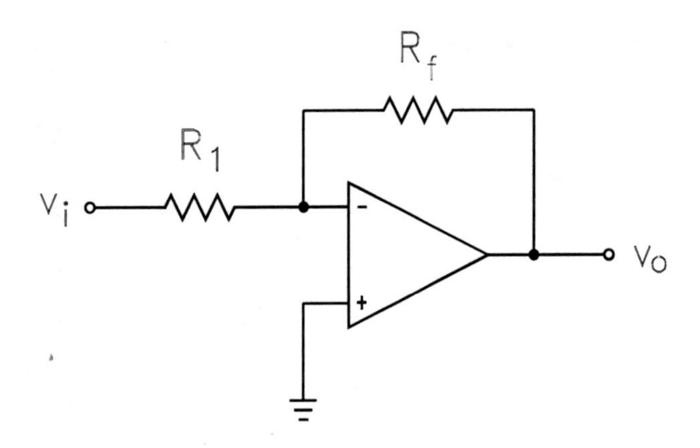

Figure 3.2: The inverting op-amp amplifier.

The inverting op-amp amplifier is shown in Fig. 3.2. The voltage gain is given by

$$\frac{v_o}{v_i} = -\frac{R_f}{R_1} \tag{3.2}$$

where R_f, the feedback resistor, is chosen to be larger than R_1, the input resistor. The output and input voltage signals are 180° out of phase (which is why it is known as an inverting amplifier) and the amplitude of the output is R_f/R_1 larger than the input. Since the open loop gain of the op-amp itself is infinite, the voltage at both the inverting and noninverting input of the op-amp are the same; the noninverting input is grounded and the inverting input is at a "virtual ground". The input impedance is R_1.

The amplifier shown in Fig. 3.3 is the noninverting op-amp amplifier. The voltage gain of this amplifier is given by

$$\frac{v_o}{v_i} = 1 + \frac{R_f}{R_1} \tag{3.3}$$

The input and output voltage signals are in phase (which is why it is known as the noninverting amplifier). The input impedance is R_i which can be selected to be a rather large value which makes the input impedance of the noninverting amplifier much larger, in general, than the input impedance of the inverting amplifier. If $R_f = 0$, then this amplifier becomes known as a unity gain buffer.

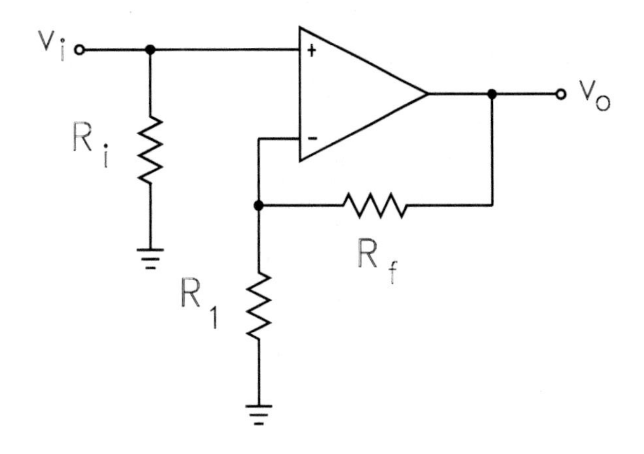

Figure 3.3: The noninverting op-amp amplifier.

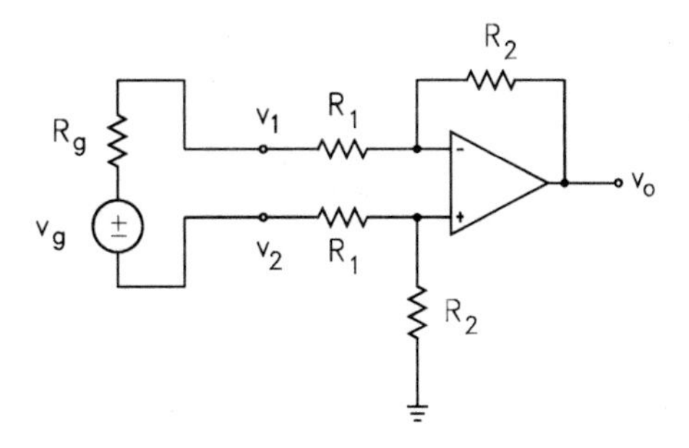

Figure 3.4: Op-amp differential amplifier.

The amplifier shown in Fig. 3.4 is known as an op-amp difference or differential amplifier. The source is shown as a voltage source v_g in series with a resistor R_g which is the resistance of the voltage source, e.g. a transducer. The output is given by

$$v_o = -\frac{R_2}{R_1}(v_1 - v_2) = \frac{2R_1}{R_g + 2R_1}\left[-\frac{R_2}{R_1}\right]v_g \tag{3.4}$$

which is a function of the difference of v_1 and v_2. This type of amplifier is used when low level signals from a transducer such as a strain gage or a microphone are to be amplified. In such applications, the differential amplifier is superior to either the inverting or noninverting amplifiers in rejecting unwanted noise sources known as common mode noise.

The expression for the voltage gain of the differential amplifier, Eq. 3.4, shows that R_1 should be large compared to R_g or the voltage gain well be significantly degraded, an effect known as source, generator, or transducer loading. This condition can be easily met with transducers such as strain gages which have low source impedances. Other transducers, such as ECG (electrocardiogram) electrodes, have high source impedances which requires that a refinement of the differential amplifier known as the instrumentation amplifier be used.

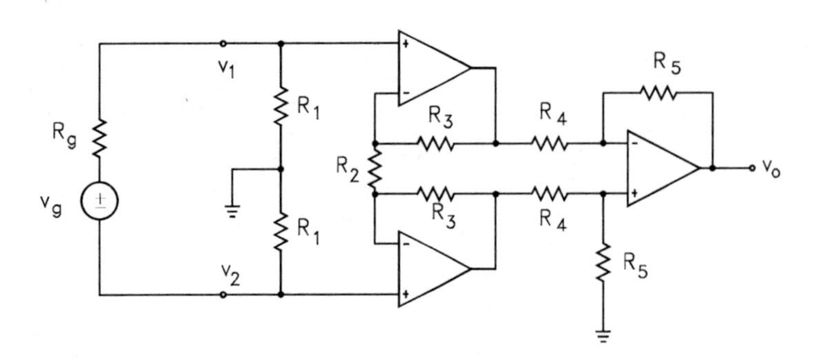

Figure 3.5: Op-amp instrumentation amplifier.

The circuit shown in Fig. 3.5 is known as the instrumentation amplifier. The purpose of the op-amps A_1 and A_2 is to provide a high input impedance. The output voltage is given by

$$v_o = -\frac{R_5}{R_4}(v_3 - v_4) = -\frac{R_5}{R_4}\left[1 + \frac{2R_3}{R_2}\right](v_1 - v_2) \tag{3.5}$$

The output can be expressed as a function of v_g by

$$v_o = -\frac{R_5}{R_4}\left[1 + \frac{2R_3}{R_2}\right]\frac{2R_1}{R_g + 2R_1}v_g \tag{3.6}$$

For this amplifier, R_1 can be picked to be much larger than R_g so that the transducer is not appreciably loaded.

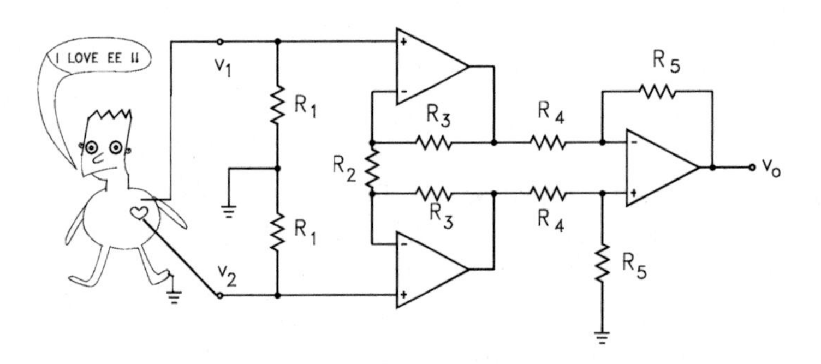

Figure 3.6: Measurement of ECG with an instrumentation amplifier.

A practical application of an instrumentation amplifier is shown in Fig. 3.6. An ECG is to be measured for the individual on the left. Transducers known as electrodes are attached above and below the heart and the toe is grounded. The electrical signals produced by the heart are in the millivolt range. Corrupting this measurement is 60 Hz interference which is in the 100 millivolt range and both electrodes also produce an unwanted DC voltage in the range of several hundred millivolts. The instrumentation amplifier cancels both the 60 Hz interference and DC offset voltage produced by the transducer. Only the ECG signal is present at the output of the instrumentation amplifier which may then be connected to the input of an oscilloscope, strip chart recorder, or an analog to digital converter for conversion to digital form which could then be processed by a computer.

Ideally, neither the difference amplifier or the instrumentation amplifier would respond to a common mode signal, i.e. if the two inputs were connected together, the output would be zero. Due to mismatches in resistor values and components inside physical op-amps there will be a small nonzero output voltage when the inputs are identical. A measure of an actual amplifier's ability to reject common mode signals is the common mode rejection ratio which is defined as

$$CMRR = \left|\frac{DG}{CG}\right| \tag{3.7}$$

the ratio of the differential mode gain to the common mode gain. The differential mode gain is measured and then the common mode gain and the CMRR can be calculated from Eq. 3.7. This is usually a large number and is often expressed in decibels ($20 \log_{10} CMRR$).

3.3 Procedure

3.3.1 Pin Connections for Op-Amps

The op-amps that will be used in this experiment are the 741 and the TL071. These op-amps are contained in a physical package known as a dual-in-line (DIP); it resembles a small gray or black bug with eight legs. The active components are encased in a plastic or ceramic package and eight metal leads (known as pins) for connection to electronic circuits. This package can be easily inserted in a breadboard so that pins are on either side of the trough.

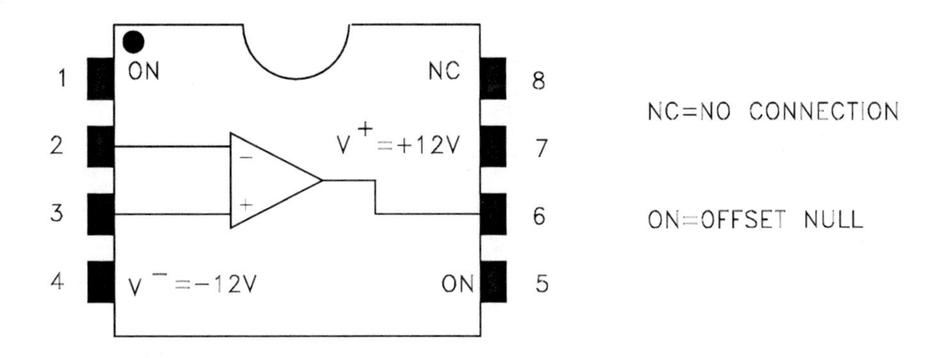

Figure 3.7: Pin connections for 741 type op-amp.

The pin connections (aka pinouts) for the both the 741 and TL071 are shown in Fig. 3.7. Pin 8 is never connected to anything. Pins 1 and 5 are used to null the DC offset produced at the output of these op-amps which will not be necessary for any of the circuits examined in this course. Pin 7 must be connected to the positive DC power supply and pin 4 must be connected to the negative DC power supply. Pin 6 is the output and pin 3 is the noninverting input and pin 2 is the inverting input. Either of these op-amps will work with DC voltages in the range 8 V to 18 V; the value that will be used in these experiments is +12 V.

3.3.2 Power Supply Adjustment

Turn on the **Tektronix DMM4040 Digital Multimeter** (DMM) and set the Digital Multimeter to measure a DC voltage (DC V) which is the mode that it boots up in. Connect one lead to the yellow binding post in the upper right of the **CADET** and the other to the black binding post below it. Vary the position of the $+V$ control to the left of the binding posts until the voltage indicates 12 V. Move the red lead of the DMM to the blue binding post and vary the $-V$ control until the voltage indicated is -12 V. Turn the DMM off. (The black binding post is known as ground because it is connected through one of the wires in the AC power cord for the **CADET** to a wire running deep into the third planet from the sun. It is used as a reference for 0 V.)

Make connections to establish buses for the positive and negative DC power supplies and ground as shown in Fig. 3.8. The top four rows on the **CADET** are horizontal power supply buses; they are connected internally to the four binding posts and, therefore, it is not necessary to use wire jumpers from the binding posts to the top four rows. To make connections to ICs wire jumpers must be used to make connections from the appropriate power supply bus on one of the top four rows to buses on the lower breadboard where the circuits will be assembled. The vertical buses on the breadboard will be used for $+12$ V, -12 V, and ground as shown in Fig. 3.8. These are to be separate buses and at no time should any of these three be directly connected to another. Note that the vertical buses are broken half way down and wire jumpers must be used if circuits are to be assembled on the lower half of the breadboard. Note that there are three identical breadboard sockets (only one is shown in Fig. 3.8) and that the drawing is not to scale.

3.3.3 Safety

Components should be changed only when the power supply is turned off. The AC switch in the upper left hand corner of the **CADET** should always be turned off when the circuit is being assembled or the components are being changed.

Should a component such as a resistor or IC over heat and emit smoke, turn the power supply off and then, and only then, take corrective action. A component which has emitted smoke should not be touched until it has had time to cool off (a few minutes). Do NOT touch a hot circuit component or a painful burn may occur.

3.3.4 Inverting Amplifier

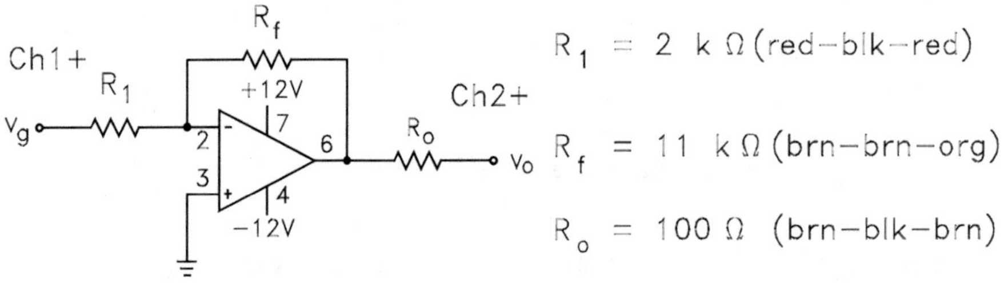

R_1 = 2 kΩ (red−blk−red)

R_f = 11 kΩ (brn−brn−org)

R_o = 100 Ω (brn−blk−brn)

Figure 3.9: Inverting amplifier.

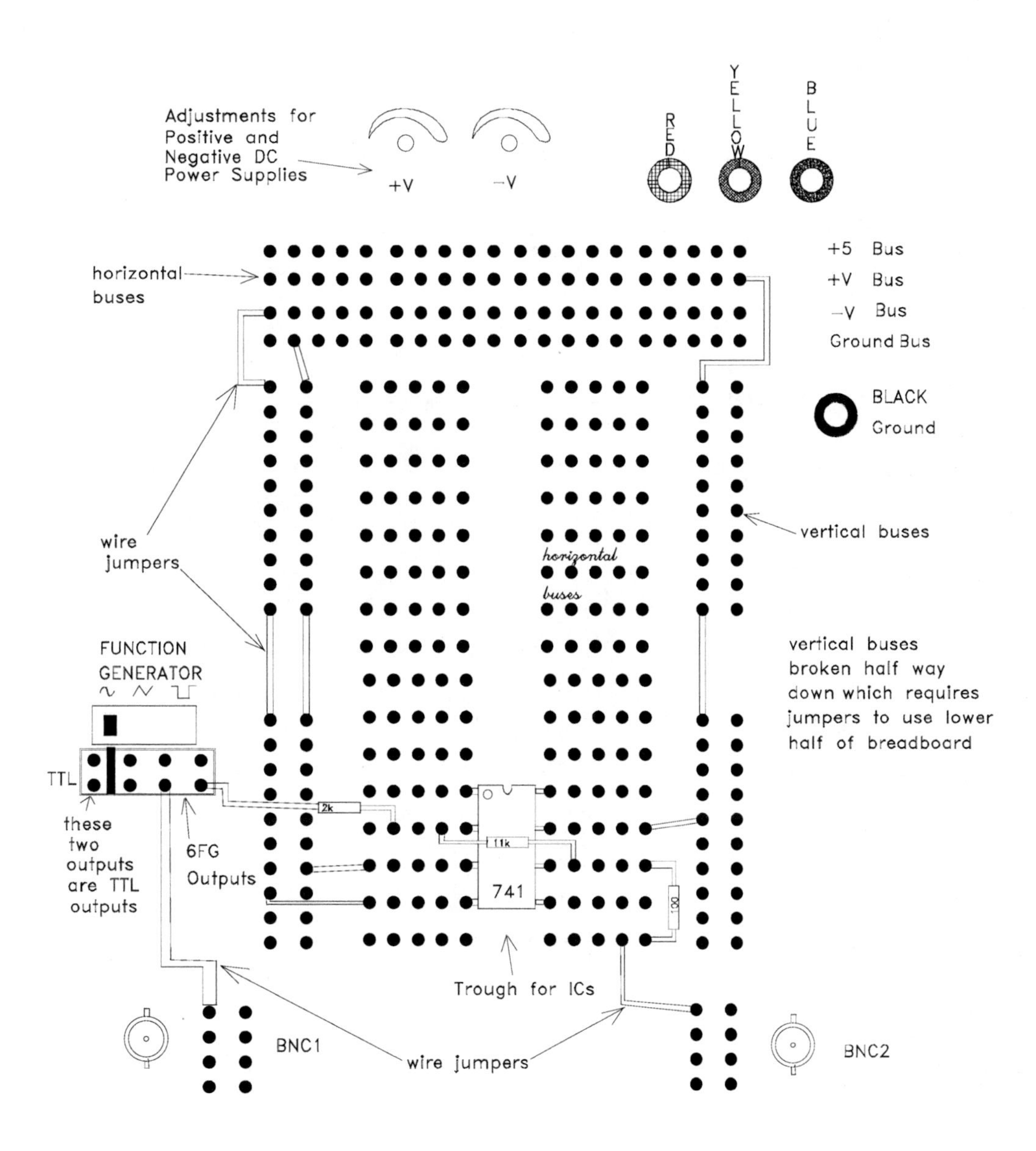

Figure 3.8: DC power supply buses for op-amp experiments.

Assemble the circuit shown in Fig. 3.9 on the **CADET** as shown in Fig. 3.8. Use the 741 op-amp. Connect $CH1$ of the scope to the input of the amplifier and $CH2$ to the output of the amplifier. (Use the BNC connectors at the lower part of the **CADET** to make the connection from the circuit assembled on the **CADET** and the input to the scope.). Set the frequency of the Function Generator on the **CADET** to 1 kHz and the amplitude to 1 V. Use the sine output of the Function Generator. Turn on the oscilloscope. Press **Defaut Setup**. Turn Ch2 on and press **Autoset**. Turn on the measurements to measure the frequency of the Ch1 input, the peak-to-peak voltages on Ch1 and Ch2. Print the display. Demonstrate the properly functioning circuit to the laboratory instructor.

What is the experimental voltage gain? ______________________________

What is the theoretical voltage gain? ______________________________

What is the phase shift between the input and output? ______________________________

Laboratory Instructor Verification ______________________________

3.3.5 Noninverting Amplifier

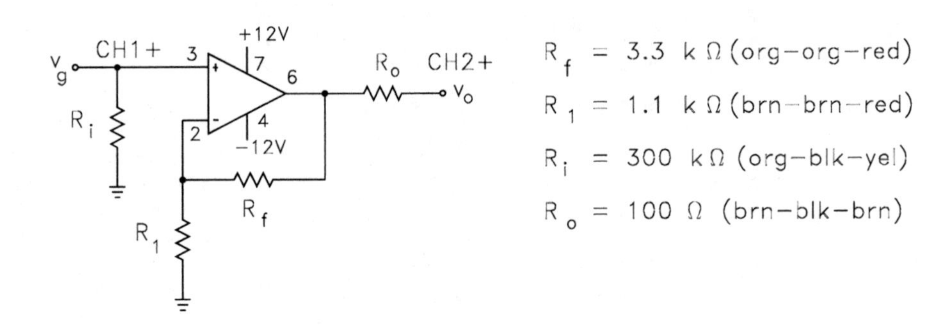

Figure 3.10: Noninverting amplifier.

Assemble the circuit shown in Fig. 3.10. Use the 741 op-amp. Use the same settings for the Function Generator as the previous step. Connect the function generator output as v_g. Press **Autoset** on the scope. Print the display. Demonstrate the properly functioning circuit to the Laboratory Instructor.

What is the experimental voltage gain? ______________________________

What is the theoretical voltage gain? ______________________________

What is the phase shift between the input and output? ______________________________

Laboratory Instructor Verification ______________________________

3.3.6 Differential Amplifier

Differential Mode Gain

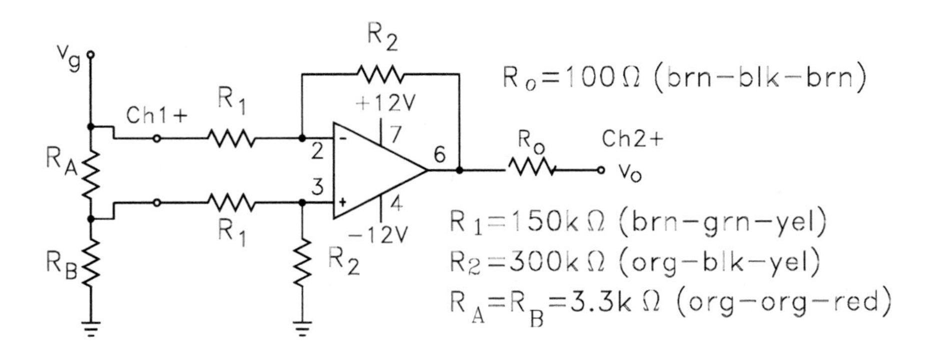

Figure 3.11: Differential amplifier.

Assemble the circuit shown in Fig. 3.11. Use the 741 op-amp. Use the same Function Generator settings as the previous two steps. Connect the output of the circuit to $CH2$ and the input to $CH1$. Press **Autoset**. Print the display or sketch the waveforms on the enclosed graph paper. Turn on the **Tektronix DMM4040 Digital Multimeter** (DMM) and set the Digital Multimeter to measure a AC voltage (AC V). Measure the input voltage by measuring the ac voltage across R_A with the DMM. Measure the ac voltage from the output node to ground with the DMM. Demonstrate the properly functioning circuit to the laboratory instructor. The differential voltage gain is output voltage divided by the differential input voltage, i.e. the peak value of the output divided by the difference in the peak values of the two inputs.

What is the experimental differential voltage gain? ______________________________

What is the theoretical differential voltage gain? ______________________________

What is the phase shift between the input and output? ______________________________

Laboratory Instructor Verification ______________________________

Common Mode Gain

Replace the resistor R_A with a short circuit; this makes the two inputs the same. Use the Digital Multimeter to measure the rms value of the output voltage. Set the DMM to AC (AC V) and connect the red lead to the output of the amplifier and the black lead to ground. Measure the rms of the input voltage. The common mode gain is the measured value of rms value of the output divided by the rms value of the input.

What is the common mode gain? ______________________________

What is the common mode rejection ratio in decibels? ______________________________

3.3.7 Instrumentation Amplifier

Differential Mode Gain

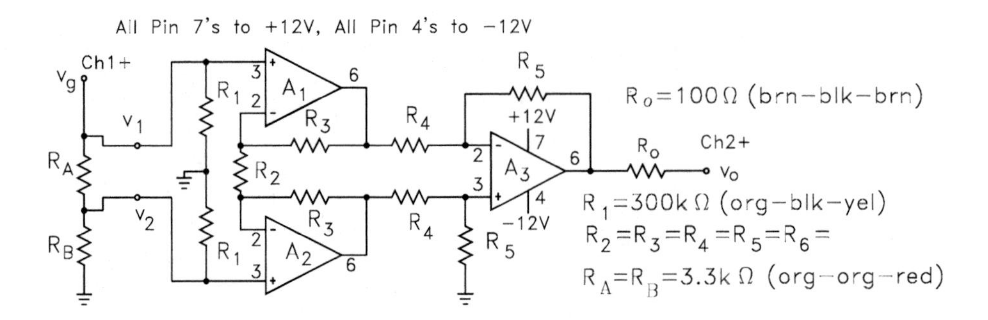

Figure 3.12: Instrumentation amplifier.

Assemble the circuit shown in Fig. 3.12. Use the TL071 op-amps for A1 and A2 and the 741 for A3. Use the same Function Generator settings as the previous two steps. Print the display or sketch the waveforms on the enclosed graph paper. Measure the input voltage by measuring the ac voltage across R_A with the DMM. Measure the ac voltage from the output node to ground with the DMM. Demonstrate the properly functioning circuit to the laboratory instructor. The differential voltage gain is output voltage divided by the differential input voltage, i.e. the peak value of the output divided by the difference in the peak values of the two inputs. Demonstrate the properly functioning circuit to the laboratory instructor. The differential voltage gain is the peak value of the output divided by the difference in the peak values of the two inputs.

What is the experimental differential voltage gain? ____________________

What is the theoretical differential voltage gain? ____________________

What is the phase shift between the input and output? ____________________

Laboratory Instructor Verification ____________________

Common Mode Gain

Replace the resistor R_A with a short circuit; this makes the two inputs the same. Use the Digital Multimeter to measure the rms value of the output voltage. Set the DMM to AC (AC V) and connect the red lead to the output of the amplifier and the black lead to ground. Measure the rms value of the input. The common mode gain is the measured value of rms value of the output divided by the rms value of the input.

What is the common mode gain? ____________________

What is the common mode rejection ratio in decibels? ____________________

3.4 Laboratory Report

Turn off the DMM and scope. Turn in all sketches, printouts, and the answers to the questions posed in the procedure section as well as any supplementary questions which may have been posed by the laboratory instructor.

Title:____________________________

Figure No.________________________

VOLTS/DIV

Ch1 ________
Ch2 ________
Ch3 ________
Ch4 ________

TIME/DIV
MAIN
____
DELAYED
____
DELAY
____

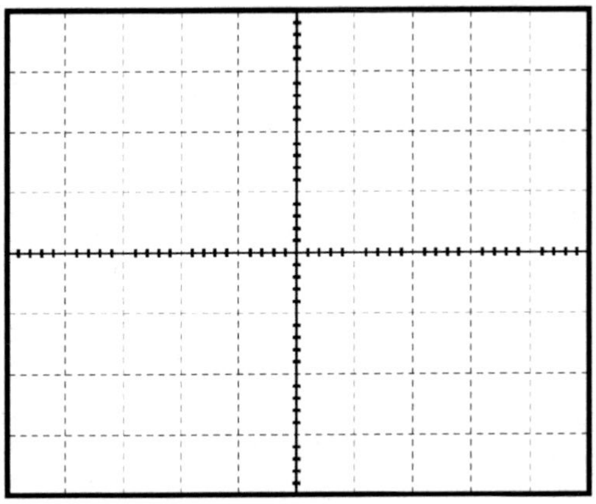

Coupling
AC ☐
DC ☐
Ch 2 Inv ☐
BW Limit ☐

Title:____________________________

Figure No.________________________

VOLTS/DIV

Ch1 ________
Ch2 ________
Ch3 ________
Ch4 ________

TIME/DIV
MAIN
____
DELAYED
____
DELAY
____

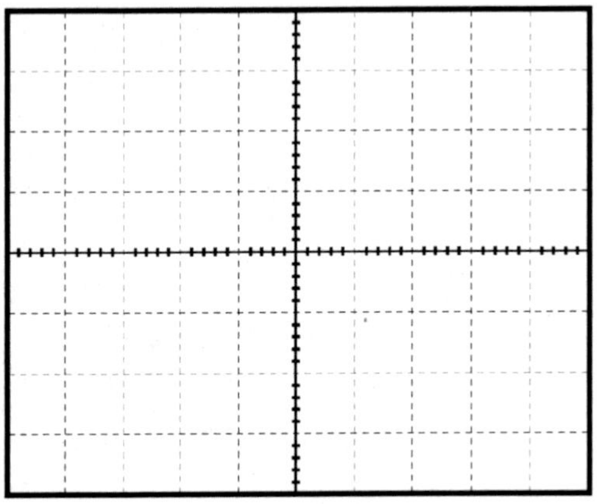

Coupling
AC ☐
DC ☐
Ch 2 Inv ☐
BW Limit ☐

Figure 3.13:

Title:____________________

Figure No.____________________

VOLTS/DIV

Ch1 ________
Ch2 ________
Ch3 ________
Ch4 ________

TIME/DIV
MAIN

____

DELAYED

____

DELAY

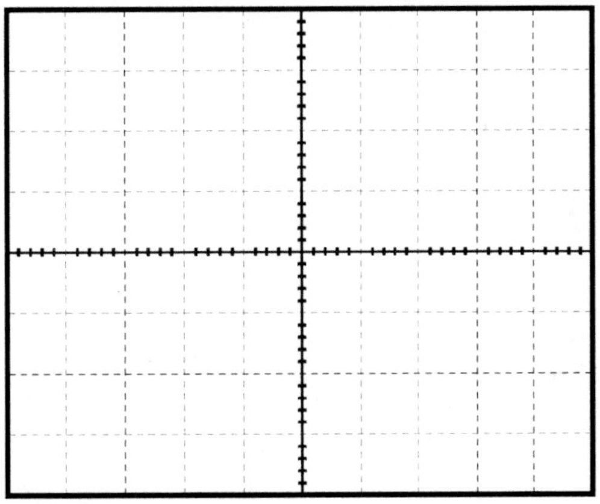

____

Coupling
AC ☐
DC ☐
Ch 2 Inv ☐
BW Limit ☐

Title:____________________

Figure No.____________________

VOLTS/DIV

Ch1 ________
Ch2 ________
Ch3 ________
Ch4 ________

TIME/DIV
MAIN

____

DELAYED

____

DELAY

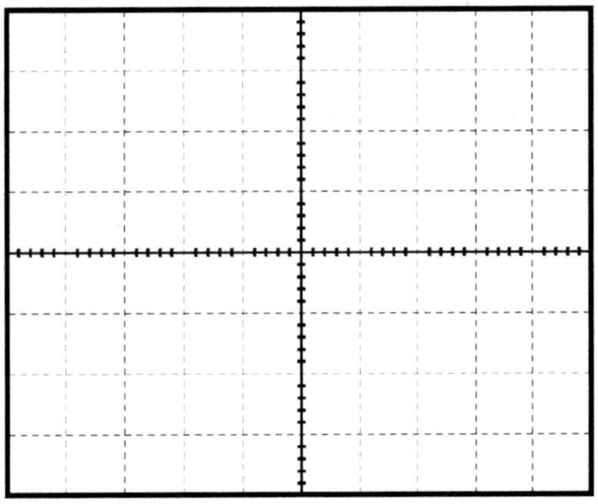

____

Coupling
AC ☐
DC ☐
Ch 2 Inv ☐
BW Limit ☐

Figure 3.14:

Chapter 4

Op-Amp Filters

4.1 Objective

The objective of this experiment is to experimentally examine first- and second-order op-amp active filters. Low-pass, band-pass, and high-pass filters will be examined.

4.2 Theory

Filters are common elements in electronic systems. They are used to pass frequencies in certain bands of the frequency spectrum while rejecting frequencies outside of these bands. Without filters electronic communication would be impossible and civilization would revert back to the Dark Ages. Shown in Fig. 4.1 is the frequency transfer characteristic or frequency transfer function of an ideal low-pass, band-pass, and high-pass filter.

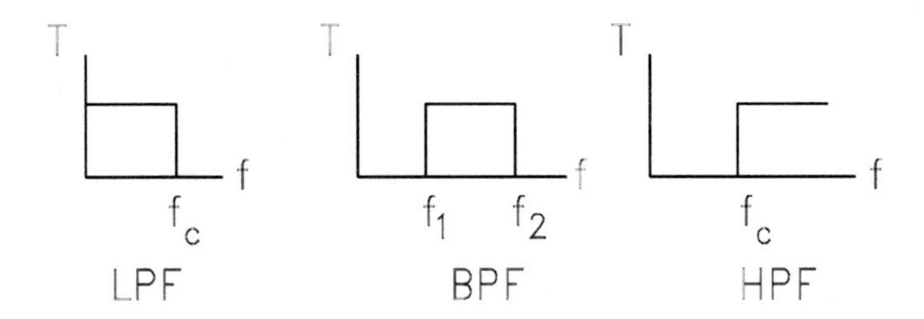

Figure 4.1: Frequency transfer characteristic of ideal low-, band-, and high-pass filter.

The ideal low-pass filter passes all frequencies that are less than the frequency f_c which is known as the cutoff frequency. The frequency band $0 < f < f_c$ is known as the pass-band while the frequency band $f_c < f < \infty$ is known as the stop band. The gain in the pass-band is unity while the gain in the stop-band is zero.

The ideal high-pass filter performs the inverse function of the ideal low-pass filter. It passes with a gain of unity all frequencies greater than f_c and stops all frequencies less than f_c. The stop-band is $0 < f < f_c$ and the pass-band is $f_c < f < \infty$.

The ideal band-pass filter passes frequencies in a certain band of the frequency spectrum and stops frequencies outside of this band. The width of the band is known as the bandwidth of the filter. For the band-pass filter shown in Fig. 4.1, the bandwidth is $BW = \Delta f = f_2 - f_1$, the pass-band is $f_1 < f < f_2$, and the stop-band is $0 < f < f_1$ and $f_2 < f < \infty$. Such a filter could be implemented by cascading an ideal low-pass filter with a cutoff frequency of f_2 with an ideal high-pass filter with a cutoff frequency of f_1.

Ideal filters do not exist outside of textbooks and the world of academe. Any real filter cannot pass from the pass-band to the stop-band as sharply as those shown in Fig. 4.1. The flat frequency response (amplitude of frequency transfer characteristic being constant in the pass or stop bands) is also not possible to realize with a real filter. All real filters exhibit some variation of the voltage gain with frequency in both the pass- and stop-bands. If the filter is a linear system, the voltage transfer characteristic can be expressed as a complex transfer function with the order of the filter being the order of the transfer function. As the order of the filter becomes larger, the filter's transfer characteristic more nearly resembles that of an ideal filter. This experiment will examine the amplitude frequency response of the first- and second-order low- and high-pass filters as well as the second-order band-pass filter (the first-order band-pass filter does not exist). The active filters will be implemented with op-amps as the active components.

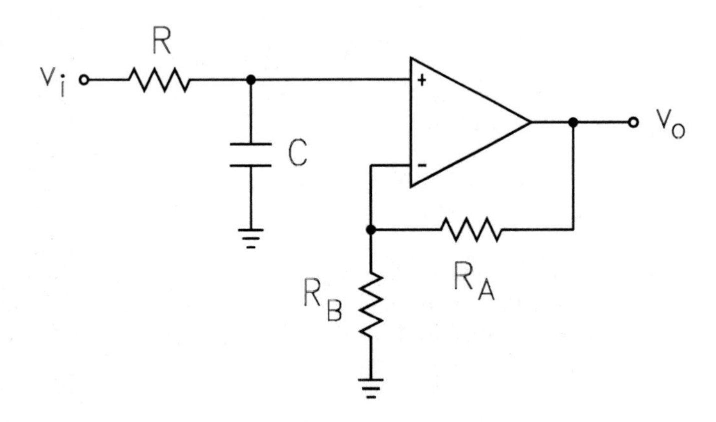

Figure 4.2: First-order low-pass filter.

The first-order low-pass filter is shown in Fig. 4.2. The transfer function is given by

$$T(s) = \frac{V_o}{V_i} = K\frac{1}{1 + \dfrac{s}{\omega_o}} \tag{4.1}$$

where

$$K = 1 + \frac{R_A}{R_B} \tag{4.2}$$

$$\omega_o = 2\pi f_o = \frac{1}{RC} \tag{4.3}$$

and, of course, $s = j\omega$. The DC gain of this circuit is K. As the frequency increases, the gain (the magnitude of the complex quantity T) decreases monotonically with the frequency. There is no magic frequency at which the filter switches from a pass-band to a stop-band. The cutoff frequency is taken as $f_c = f_o$. At this frequency the gain is 0.707 of the DC gain, K.

Gains are commonly expressed in decibels

$$| T_{dB} | = 20 \ \log_{10} |T(s)| \, |_{s=j\omega} \tag{4.4}$$

so that variations in the output signal level can be expressed on a log scale. For low frequencies ($f << f_c$) the gain in dB would be 20 $\log_{10}(K)$ while at large frequencies ($f >> f_o$) the gain would decrease by 20 dB for each increase in the frequency by a factor of 10 (this change in the frequency by a factor of 10 is known as a decade). If the gain were plotted on semilogarithmic graph paper (frequency on log axis and gain in dB on the linear axis) it would appear to consists of straight lines for frequencies well removed from f_o. The low frequency straight line (known as the low frequency asymptote) would be a horizontal line parallel to the frequency axis while the high frequency straight line (known as the high frequency asymptote) would have a slope of -20 *db/decade* and intersect the low frequency asymptote at $f = f_o$. At the frequency f_o the gain is -3 dB and for this reason another name for f_c is the minus 3 dB frequency.

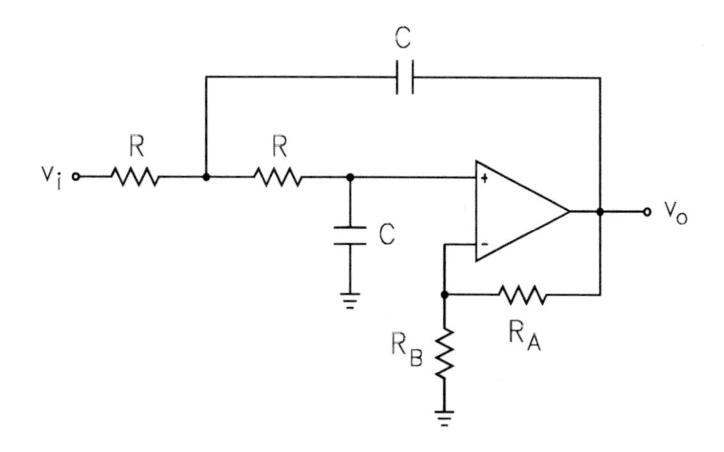

Figure 4.3: Second-order Sallen-Key low-pass filter.

Many circuit topologies will implement a second order low-pass filter. A second-order low-pass filter known as the Sallen Key filter is shown in Fig. 4.3. The complex voltage transfer function for this filter is

$$T(s) = \frac{V_o}{V_i} = K\frac{1}{1+\frac{1}{Q}\left(\frac{s}{\omega_o}\right)+\left(\frac{s}{\omega_o}\right)^2} \tag{4.5}$$

where

$$\omega_o = 2\pi f_o = \frac{1}{RC} \tag{4.6}$$

$$K = 1 + \frac{R_A}{R_B} \tag{4.7}$$

$$Q = \frac{1}{3-K} \tag{4.8}$$

and K is the DC gain of the filter and Q is known as the quality factor of the filter. The cutoff frequency of the filter is given by

$$f_c = x\ f_o \tag{4.9}$$

where the factor x is

$$x^2 = 1 - \frac{1}{2Q^2} + \sqrt{1+\left(1-\frac{1}{2Q^2}\right)^2} \tag{4.10}$$

which means that f_c is larger than f_o unless $Q = 1/\sqrt{2}$ when they are equal.

If the gain of the second order low-pass filter is plotted as a function of frequency on semilogarithmic graph paper, the low frequency asymptote would be a straight line parallel to the frequency axis with a value of 20 $\log_{10}(K)$. The high frequency asymptote would be a straight line with a slope of -40 *db/dec*. The high and low frequency asymptotes would intersect at $f = f_o$. At the frequency f_o the value of the gain is QK.

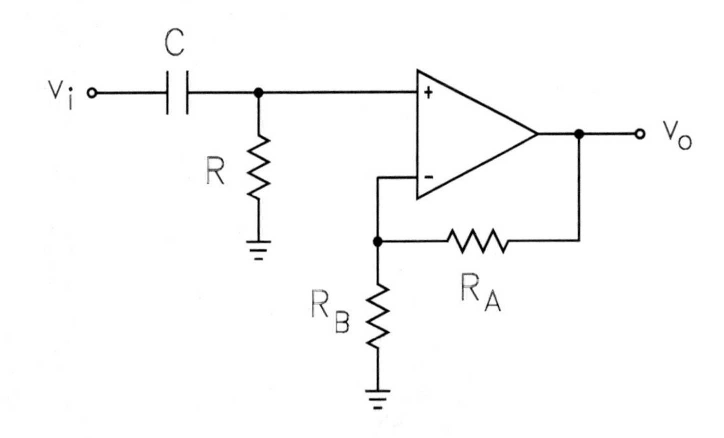

Figure 4.4: First-order high-pass filter.

The first order high-pass filter is shown in Fig. 4.4. The complex transfer function for this filter is given by

$$T(s) = K\frac{\frac{s}{\omega_o}}{\frac{s}{\omega_o} + 1} \tag{4.11}$$

where K is the high frequency gain which is given by

$$K = 1 + \frac{R_A}{R_B} \tag{4.12}$$

The cutoff frequency is given by

$$f_c = f_o = \frac{1}{2\pi RC} \tag{4.13}$$

and is the frequency at which the gain is down by 3 dB from the high frequency gain. If the gain in dB is plotted on semilogarithmic graph paper, it will appear to consist of straight line segments for frequencies well removed from f_c. The high frequency asymptote will be a straight line parallel to the frequency axis with a value of 20 $\log_{10}(K)$. The low frequency asymptote will be a straight line with a slope of $+20$ *db/dec*. The high and low frequency asymptote intersect at the frequency f_o.

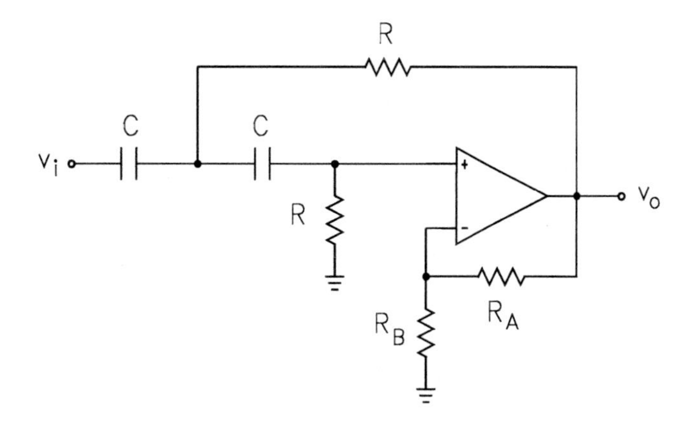

Figure 4.5: Second-order Sallen-Key high-pass filter.

The second-order Sallen Key high-pass filter is shown in Fig. 4.5. The complex transfer function for this filter is given by

$$T(s) = \frac{V_o}{V_i} = K \frac{\left(\frac{s}{\omega_o}\right)^2}{1 + \frac{1}{Q}\left(\frac{s}{\omega_o}\right) + \left(\frac{s}{\omega_o}\right)^2} \tag{4.14}$$

where K is the high-frequency gain given by

$$K = 1 + \frac{R_A}{R_B} \tag{4.15}$$

$$\omega_o = \frac{1}{RC} \tag{4.16}$$

and

$$Q = \frac{1}{3 - K} \tag{4.17}$$

is known as the quality factor for the filter. The cutoff or minus 3 dB frequency is given by $f_c = f_o/x$ where

$$x^2 = 1 - \frac{1}{2Q^2} + \sqrt{1 + \left(1 - \frac{1}{2Q^2}\right)^2} \tag{4.18}$$

which means that f_c is not equal to f_o unless $Q = 1/\sqrt{2}$ and, in general, is less than f_o.

If the gain is plotted as a function of frequency for the second-order low-pass filter on semilogarithmic graph paper, the plot will appear to be straight lines for frequencies well removed from f_o. The high frequency asymptote will be a straight line parallel to the frequency axis with a value of 20 $\log_{10}(K)$. The low asymptote will be straight line with a slope of 40 db/dec that will intersect the low frequency asymptote at the frequency $f = f_o$. At the frequency f_o the gain will be KQ.

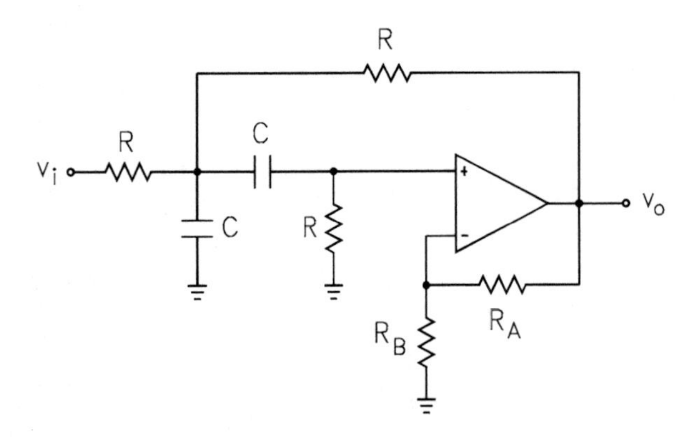

Figure 4.6: Second-order Sallen-Key bandpass filter.

The second-order Sallen Key band-pass filter is shown in Fig. 4.6. The complex transfer function for this filter is given by

$$T(s) = \frac{V_o}{V_i} = K \frac{\frac{1}{Q}\left(\frac{s}{\omega_o}\right)}{1 + \frac{1}{Q}\left(\frac{s}{\omega_o}\right) + \left(\frac{s}{\omega_o}\right)^2} \tag{4.19}$$

where f_o is the center frequency of the filter, i.e. the center of the pass-band. The gain decreases monotonically as the frequency increases or decreases from f_o. There are two minus 3 dB frequencies on either side of f_o. The bandwidth is taken as the difference of the two minus 3 dB frequencies and is given as

$$\Delta f = \frac{f_o}{Q} \tag{4.20}$$

The gain at the center frequency f_o is given by K where

$$K = \frac{1}{\frac{4}{K'} - 1} \tag{4.21}$$

and

$$K' = 1 + \frac{R_A}{R_B} \tag{4.22}$$

$$\omega_o = \frac{\sqrt{2}}{RC} \tag{4.23}$$

$$Q = \frac{1}{\sqrt{2}\left(2 - \frac{K'}{2}\right)} \tag{4.24}$$

If the gain of the second-order band-pass filter is plotted as a function of frequency on semilogarithmic graph paper, the low and high frequency asymptotes appear to be straight lines. The high frequency asymptotes has a slope of $-20\ db/dec$ while the low frequency asymptotes has a slope of 20 db/dec. The asymptotes intersect at the frequency f_o.

4.3 Procedure

The op-amp that will be used to assemble the filter circuits in this experiment is the TL071. The pinouts for this IC are given in Exp. 3.

4.3.1 Power Supply Adjustment

Turn on the **Tektronix DMM4040** Digital Multimeter (DMM) and set it to measure DC voltages (DCV). Check the positive and negative DC supplies to see if they are set to +12 and −12 Volts. Turn the DMM off.

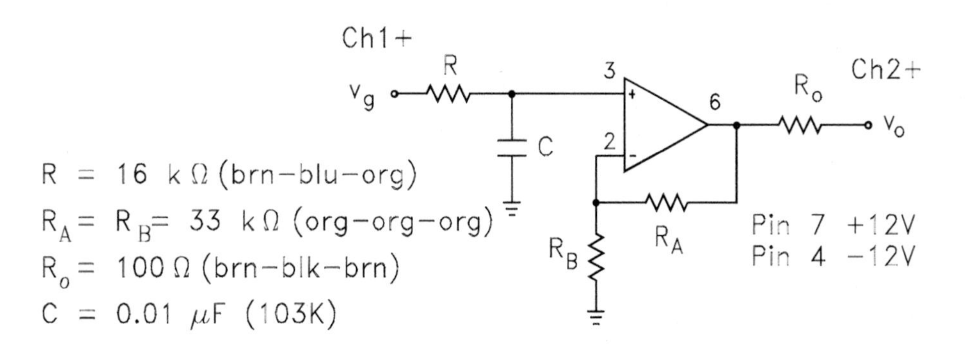

Figure 4.7: First-order low-pass filter.

4.3.2 First Order Low-Pass Filter

Assemble the circuit shown in Fig. 4.7. Turn the **Tektronix DPO3012** oscilloscope on. Connect $CH1\&2$ as shown. Press **Default Setup** and then turn Ch2 on. Press Autoset. Turn the measurements on to measure the frequency of the waveform connected to Ch1 and the peak-to-peak voltages of the waveforms connected to Ch1 and Ch2. Set the function generator on the **CADET** to produce a sine wave with a frequency of 100 Hz and a peak-to-peak value of 1 V. Measure the peak-to-peak value of the input and output waveforms that appear on $CH1\&2$ respectively. These measurements are to be repeated as the frequency of the function generator is changed to: 300 Hz, 700 Hz, 1 kHz, 10 kHz, and 100 kHz. Press **Autoset** on the oscilloscope each time the frequency of the source or generator is changed.

Frequency	V_1 *(Volts)*	V_2 *(Volts)*	*Gain* (V_2/V_1)
100 Hz			
300 Hz			
700 Hz			
1 kHz			
10 kHz			
100 kHz			

First-Order Low-Pass Filter

Plot gain ($V_o/V_g = V_2/V_1$) versus frequency data obtained on the enclosed log-log graph paper. Demonstrate the properly functioning circuit to the Laboratory Instructor.

Laboratory Instructor Verification ______________________________

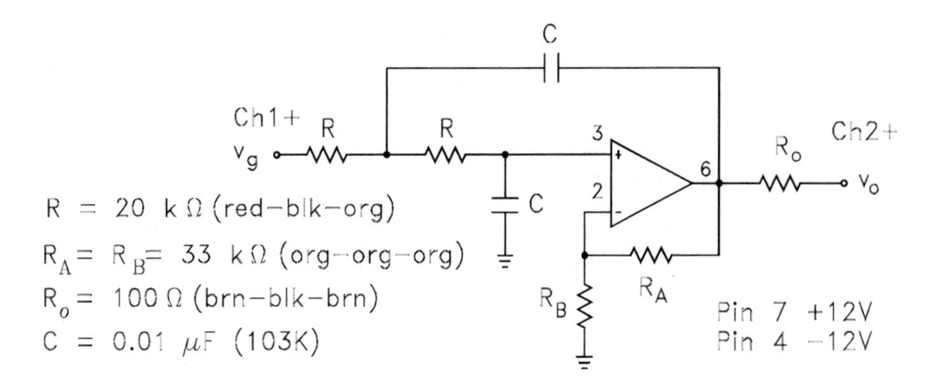

Figure 4.8: Second-order low-pass filter.

4.3.3 Second Order Low-Pass Filter

Assemble the circuit shown in Fig. 4.8. Connect $CH1\&2$ as shown. Press **Autoset**. Set the function generator on the **CADET** to produce a sine wave with a frequency of 100 Hz and a peak value of 1 V. Measure the peak-to-peak value of the input and output waveforms that appear on $CH1\&2$ respectively. These measurements are to be repeated as the frequency of the function generator is changed to: 300 Hz, 700 Hz, 1 kHz, 10 kHz, and 100 kHz. Press **Autoset** on the scope each time the frequency of the source or generator is changed.

Frequency	V_1 (*Volts*)	V_2 (*Volts*)	*Gain* (V_2/V_1)
100 Hz			
300 Hz			
700 Hz			
1 kHz			
10 kHz			
100 kHz			

Second-Order Low-Pass Filter

Plot gain ($V_o/V_g = V_2/V_1$) versus frequency data obtained on the enclosed log-log graph paper (use the same sheet of graph paper that was used for the first-order low-pass filter so that a comparison can be made between the performance of these two filters). Demonstrate the properly functioning circuit to the Laboratory Instructor.

Laboratory Instructor Verification ____

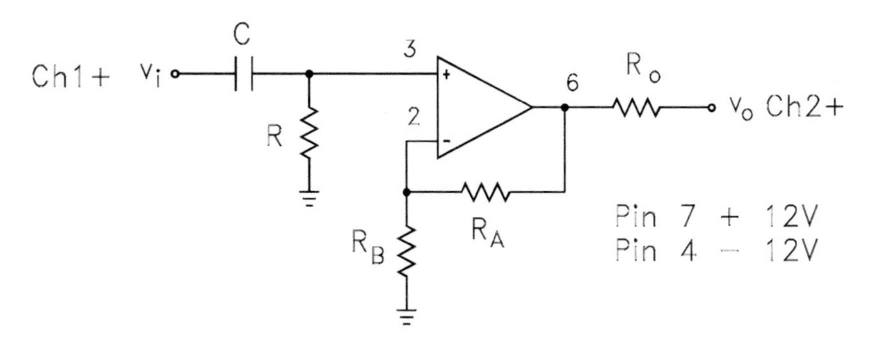

Figure 4.9: First-order high-pass filter.

4.3.4 First-Order High-Pass Filter

Assemble the circuit shown in Fig. 4.9. Connect $CH1\&2$ as shown. Set the function generator on the **CADET** to produce a sine wave with a frequency of 100 Hz and a peak value of 1 V. Press **Autoset**. Measure the peak-to-peak value of the input and output waveforms that appear on $CH1\&2$ respectively. These measurements are to be repeated as the frequency of the function generator is changed to: 1 kHz, 10 kHz, 30 kHz, and 70 kHz. Press Autoset on the oscilloscope each time the frequency of the source or generator is changed.

Frequency	V_1 (*Volts*)	V_2 (*Volts*)	*Gain* (V_2/V_1)
70 kHz			
30 kHz			
10 kHz			
1 kHz			
100 Hz			

First-Order High-Pass Filter

Plot gain ($V_o/V_g = V_2/V_1$) versus frequency data obtained on the enclosed log-log graph paper. Demonstrated the properly functioning circuit to the Laboratory Instructor.

Laboratory Instructor Verification ______________________________

R = 1.2 k Ω (brn-red-red)
R_A = R_B = 33 k Ω (org-org-org)
R_o = 100 Ω (brn-blk-brn)
C = 0.01 μF (103K)

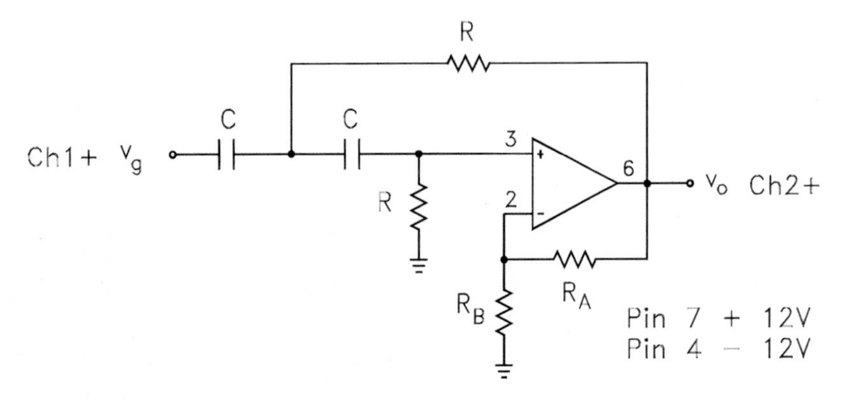

Figure 4.10: Second-order high-pass filter.

4.3.5 Second-Order High-Pass Filter

Assemble the circuit shown in Fig. 4.10. Connect $CH1\&2$ as shown. Press **Autoset**. Set the function generator on the **CADET** to produce a sine wave with a frequency of 100 Hz and a peak value of 1 V. Measure the peak-to-peak value of the input and output waveforms that appear on $CH1\&2$ respectively. These measurements are to be repeated as the frequency of the function generator is changed to: 1 kHz, 10 kHz, 30 kHz, and 70 kHz. Press **Autoset** on the scope each time the frequency of the source or generator is changed.

Frequency	V_1 (*Volts*)	V_2 (*Volts*)	*Gain* (V_2/V_1)
70 kHz			
30 kHz			
10 kHz			
1 kHz			
100 Hz			

Second-Order High-Pass Filter

Plot gain ($V_o/V_g = V_2/V_1$) versus frequency data obtained on the enclosed log-log graph paper (use the same sheet of graph paper that was used for the first-order high-pass filter so that a comparison can be made between the performance of these two filters). Demonstrated the properly functioning circuit to the Laboratory Instructor.

Laboratory Instructor Verification ______________________________

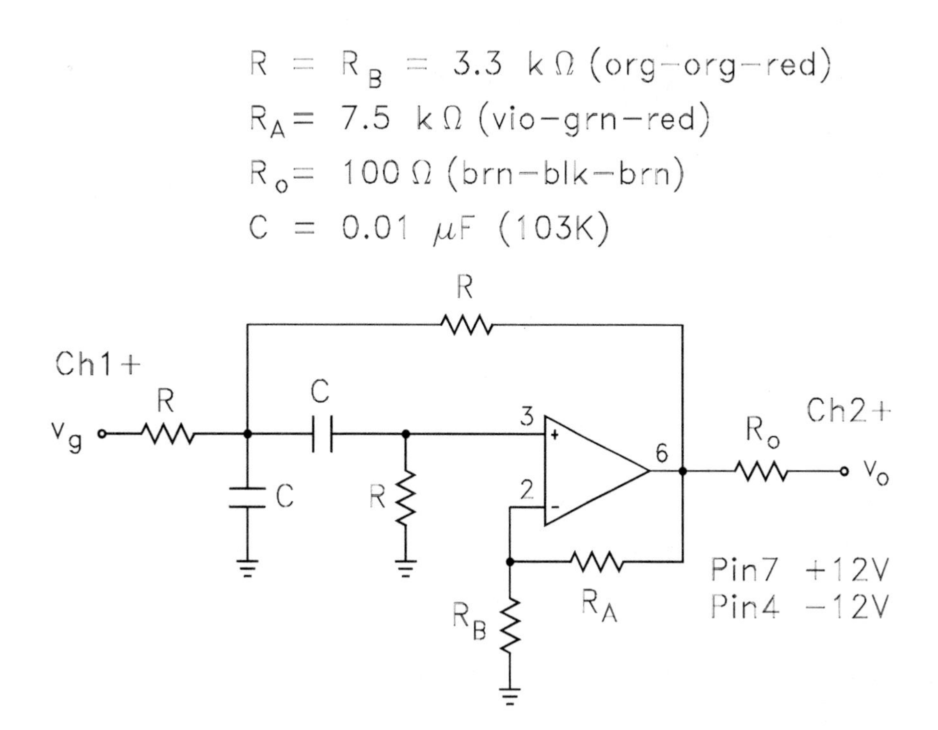

Figure 4.11: Second-order band-pass filter.

4.3.6 Second-Order Band-Pass Filter

Assemble the second-order band-pass filter shown in Fig. 4.11. Vary the frequency of the function generator until the frequency at which the output of the circuit has its maximum value. Record this frequency as f_o and the gain at this frequency. Obtain data for the gain of this circuit at frequencies: 1.3 f_o, 0.7 f_o, 10 f_o, and 0.1 f_o.

Frequency	*Hertz*	V_1 *(Volts)*	V_2 *(Volts)*	*Gain* (V_2/V_1)
1.3 f_o				
0.7 f_o				
f_o				
10 f_o				
0.1 f_o				

Second-Order Band-Pass Filter

Plot the gain versus frequency on the enclosed logarithmic graph paper.
Demonstrate the properly functioning circuit to the Laboratory Instructor.

Laboratory Instructor Verification ______________________________

4.4 Laboratory Report

From the experimental data obtained, what is the cutoff frequency for:

1. The first-order low-pass filter? ______________________________

2 The second-order low-pass filter? ______________________________

3. The first-order high-pass filter? ______________________________

4. The second-order high-pass filter? ______________________________

What is the center frequency for the band-pass filter? ______________________________

What is the bandwidth of the band-pass filter? ______________________________

Indicate on all the plots the slopes of the low and high frequency asymptotes.

Turn in all plots and the answers to the above questions as well as any supplementary questions posed by the Laboratory Instructor.

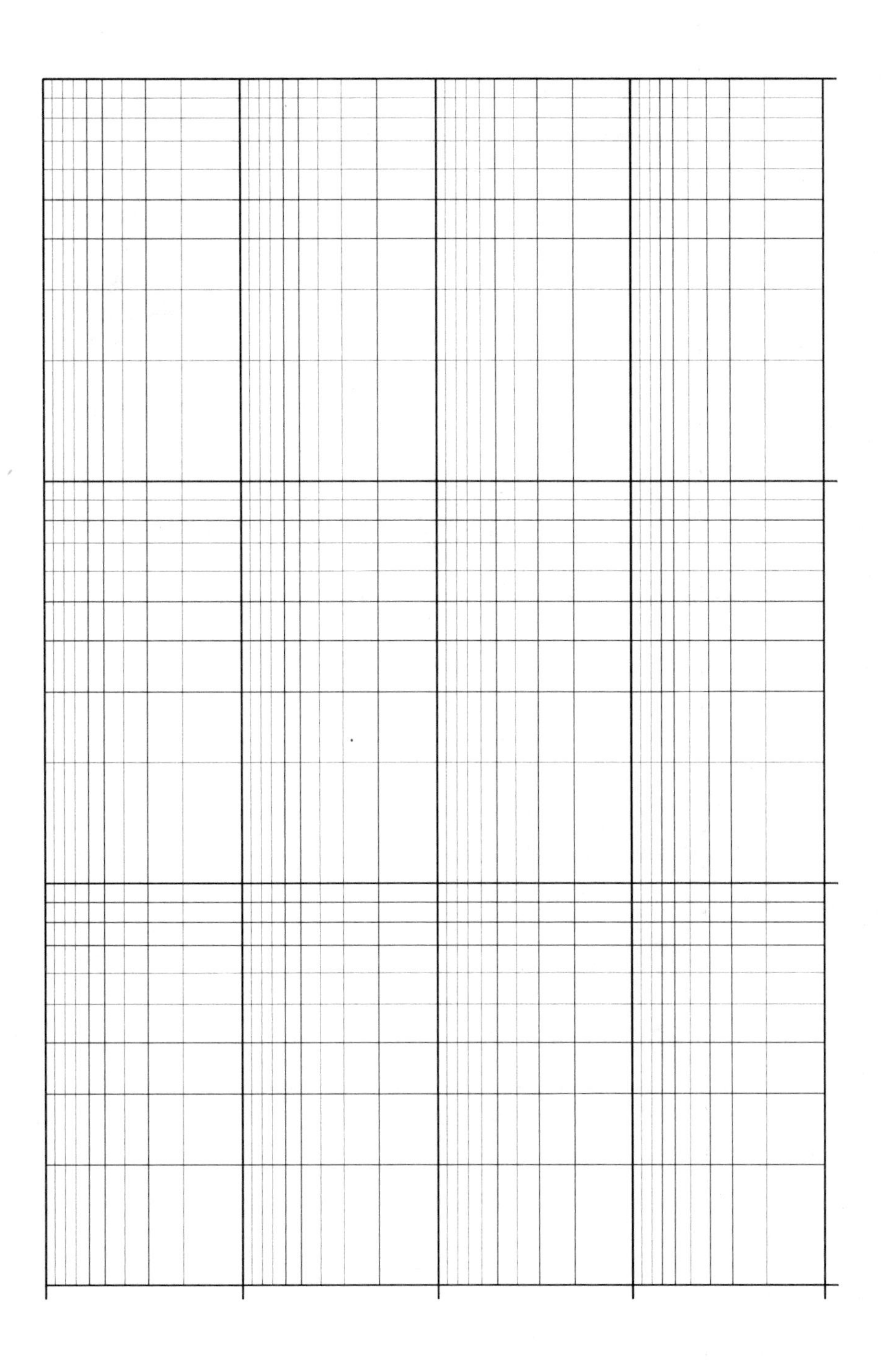

Figure 4.12:

Title:________________________________

Figure No._____________________________

VOLTS/DIV

Ch1 __________
Ch2 __________
Ch3 __________
Ch4 __________

TIME/DIV
MAIN

____

DELAYED

____

DELAY

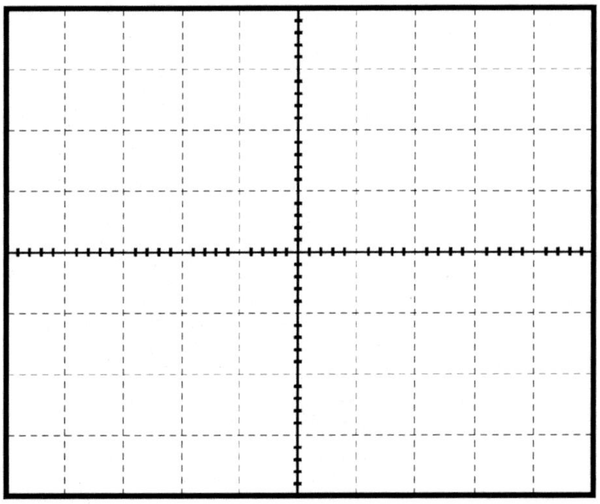

____

Coupling
AC ☐
DC ☐
Ch 2 Inv ☐
BW Limit ☐

Title:________________________________

Figure No._____________________________

VOLTS/DIV

Ch1 __________
Ch2 __________
Ch3 __________
Ch4 __________

TIME/DIV
MAIN

____

DELAYED

____

DELAY

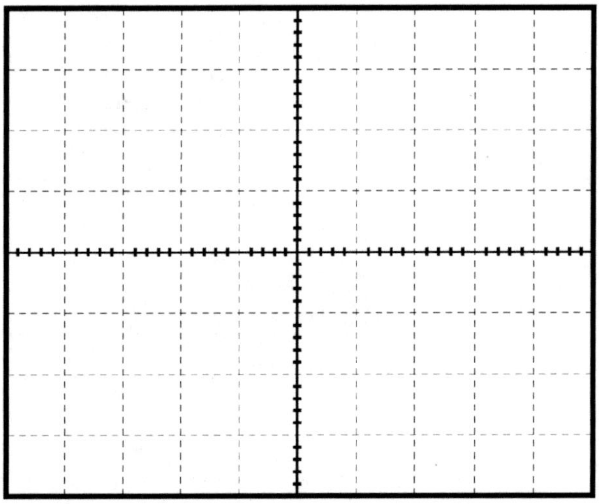

____

Coupling
AC ☐
DC ☐
Ch 2 Inv ☐
BW Limit ☐

Figure 4.13:

Chapter 5

Op-Amp Analog Computer Elements

5.1 Objective

The objective of this experiment is to experimentally examine the use of op-amps as analog computational elements. The circuits that will be examined are the integrator, the differentiator, the summer, and the state variable filter.

5.2 Theory

Historically, the initial use of op-amps were in analog computers [computers used to solve differential equations]. Indeed, the name op-amp is short for operational amplifier. Namely, these devices can be used to perform mathematical operations such as multiplication, integration, differentiation, and summation. Analog computers have been almost totally supplanted by high speed digital computers but they are still used when instantaneous solutions (real time) are needed for differential equations and the user cannot afford a super computer.

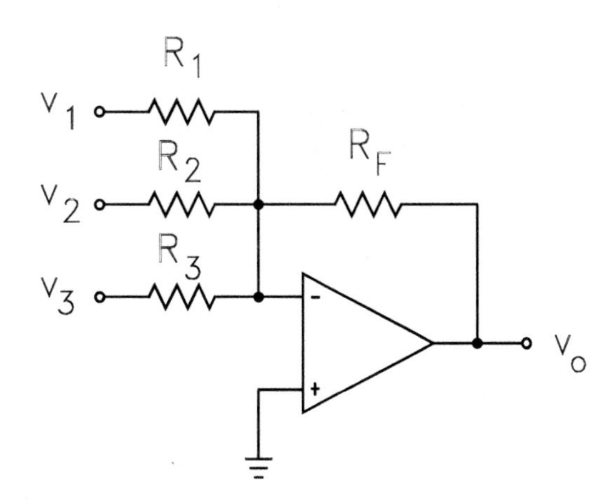

Figure 5.1: Op-amp summer.

Shown in Fig. 5.1 is an op-amp summer. The output of the summer is given by

$$v_o = -\left(\frac{R_F}{R_1}v_1 + \frac{R_F}{R_2}v_2 + \frac{R_F}{R_3}v_3\right) \tag{5.1}$$

Any number of voltages may be summed in this manner. This type of summer is known as an inverting summer.

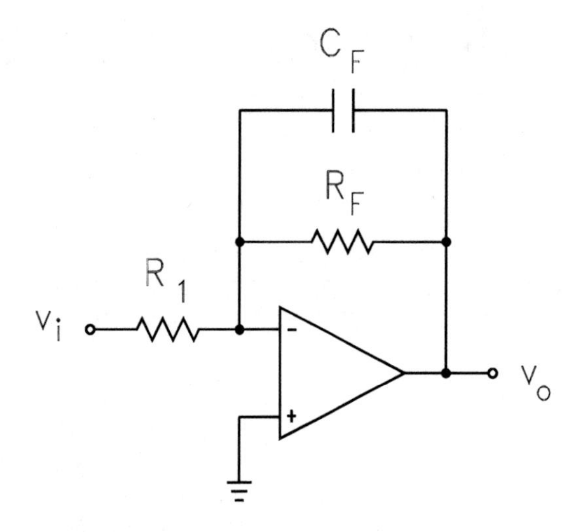

Figure 5.2: Op-amp integrator.

The circuit shown in Fig. 5.2 is known as an integrator. The complex transfer function is given by

$$T(s) = \frac{V_o}{V_i} = -\frac{1}{R_1 C_f s}\left[\frac{R_f C_f s}{1 + R_f C_f s}\right] \tag{5.2}$$

For frequencies for which $\omega >> 1/(R_f C_f)$ Eq. 5.2 becomes

$$T(s) \cong -\frac{1}{R_1 C_f s} \tag{5.3}$$

which is the transfer function of an inverting integrator with gain constant $1/(R_1 C_f)$. If the input voltage is sufficiently high in frequency this circuit's output is proportional to the integral of the input (with respect to time). Namely, the output $v_o(t)$ is given by

$$v_o(t) = -\frac{1}{R_1 C_f}\int_{-\infty}^{t} v_i(t')dt' \tag{5.4}$$

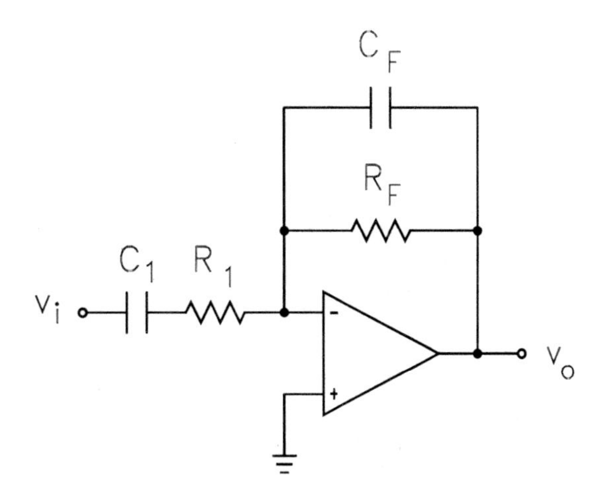

Figure 5.3: Op-amp differentiator.

The circuit shown in Fig. 5.3 is an op-amp differentiator. The complex transfer function for this circuit is given by

$$T(s) = -R_f C_1 s \left[\frac{1}{1 + R_1 C_1 s}\right] \left[\frac{1}{1 + R_f C_f s}\right] \tag{5.5}$$

If the frequency of the input satisfies $\omega << 1/(R_1 C_1)$ and $\omega << 1/(R_f C_f)$, then Eq. 5.5 becomes

$$T(s) \cong -R_f C_1 s \tag{5.6}$$

which is an inverting differentiator with a gain of $R_f C_1$ (the differentiation is with respect to time). Namely, the output is related to the input by

$$v_o(t) = -R_f C_1 \frac{dv_i(t)}{dt} \tag{5.7}$$

A linear differential equation with constant coefficients of the form

$$\frac{d^n x(t)}{dt^n} + a_{n-1}\frac{d^{n-1}x(t)}{dt^{n-1}} + \cdots + a_1 \frac{dx(t)}{dt} + x(t) = f(t) \tag{5.8}$$

can be solved for $x(t)$ and its first n derivatives if integrators and summer are available. Eq. 5.8 is simply solved for the nth derivative

$$\frac{d^n x(t)}{dt^n} = f(t) - a_{n-1}\frac{d^{n-1}x(t)}{dt^{n-1}} - \cdots - a_1 \frac{dx(t)}{dt} - x(t) \tag{5.9}$$

which is realized with a summer. The nth derivative is cascaded with n integrators which yields $x(t)$ and its n derivatives. If higher derivatives were needed a differentiator could be used.

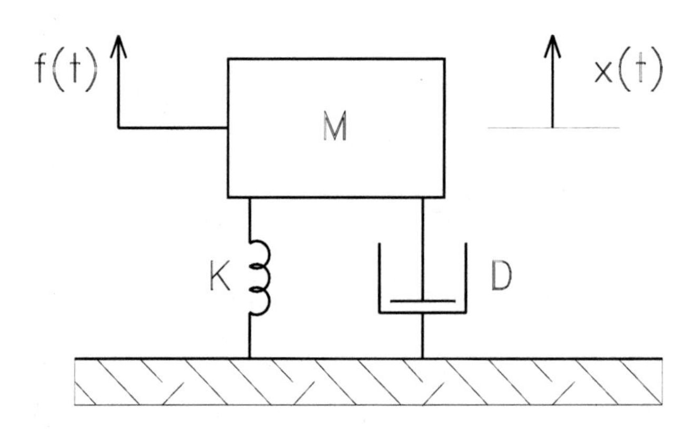

Figure 5.4: Second-order mechanical system.

Shown in Fig. 5.4 is a classical mechanical system with one degree of freedom. A force $f(t)$ is applied to a mass M which is connected to a spring with spring constant K and a dashpot with dampening coefficient D. Newton second law of motion applied to this system yields (neglecting gravity)

$$M\frac{d^2x(t)}{dt^2} = f(t) - D\frac{dx(t)}{dt} - Kx(t) \tag{5.10}$$

as the second order differential equation for the displacement $x(t)$. There are two state variables $x(t)$, the displacement or position, and $dx(t)/dt$, the velocity. This differential equation may be solved electronically using an analog computer. Two integrators would be required as well as one summer; non zero initial conditions could be realized if an initial voltage could be placed on the capacitors in the integrators. Such a circuit is also called a state variable filter.

Eq. 5.9 can be place in canonical form by dividing by the mass M which yields

$$\frac{d^2x(t)}{dt^2} + 2\,\zeta\omega_o\frac{dx(t)}{dt} + \omega_o^2 x(t) = \frac{f(t)}{M} \tag{5.11}$$

where

$$\zeta \equiv \text{dampening factor} \;=\; \frac{D}{2\sqrt{KM}} \tag{5.12}$$

$$\omega_o \equiv \text{natural frequency} \;=\; \sqrt{\frac{K}{M}} \tag{5.13}$$

are known as the dampening factor and natural frequency for this second order system.

Another parameter of interest is

$$Q = \frac{1}{2\zeta} \tag{5.14}$$

the quality factor for this second order system.

The form of the solution depends on the damping factor ζ and, of course, the type of excitation. Two types of excitation or input will be considered: step function and sinusoidal.

5.2.1 Step Function

A step function excitation is given by

$$f(t) = \begin{cases} F & t > 0 \\ 0 & t < 0 \end{cases} \tag{5.15}$$

which simply means that there is no force applied prior to $t = 0$ and at $t = 0$ a constant force of F nt is instantaneously applied.

If ζ is less than one the system is underdamped and the solution for step function excitation consists of damped or exponentially decaying sinusoidal oscillations about the final position. For ζ greater than one the system is overdamped and the solution contains exponentials which feature no oscillations. If the damping factor is one, the system is critically damped which also features no overshoot; critical damping is the boundary between the over and under damped solutions. The solution for positive t is given by

$$x(t) = \frac{F}{K}\left[1 - \frac{e^{-\zeta\omega_o t}}{\sqrt{1-\zeta^2}}\sin(\omega_o\sqrt{1-\zeta^2}t + \cos^{-1}\zeta)\right] \quad \zeta < 1 \tag{5.16}$$

$$x(t) = \frac{F}{K}[1 - (1 + \omega_o t)e^{-\omega_o t}] \quad \zeta = 1 \tag{5.17}$$

$$x(t) = \frac{F}{K}\left[1 - \frac{e^{-\zeta\omega_o t}}{\sqrt{\zeta^2 - 1}}\sinh(\omega_o\sqrt{\zeta^2 - 1}t + \cosh^{-1}\zeta)\right] \quad \zeta > 1 \tag{5.18}$$

and, of course, 0 for negative t. Typical plots of $x(t)$ versus t for step function excitation are shown in Fig. 5.5.

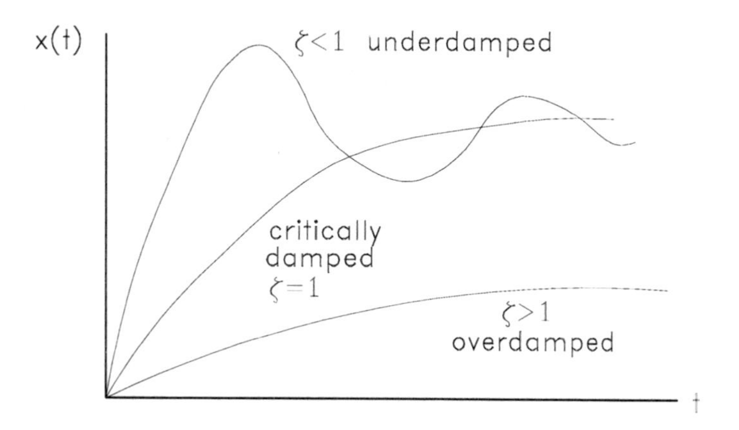

Figure 5.5: Step function response.

5.2.2 Sinusoidal Excitation

For sinusoidal excitation, i.e. $f(t) = F\ \sin(\omega t)$, the output is given by

$$x(t) = |\bar{T}(j\omega)|\sin[\omega t + \angle\bar{T}(j\omega)] \tag{5.19}$$

where $T(s)$ is the complex transfer function evaluated at $s = j\omega$. The complex transfer function is given by

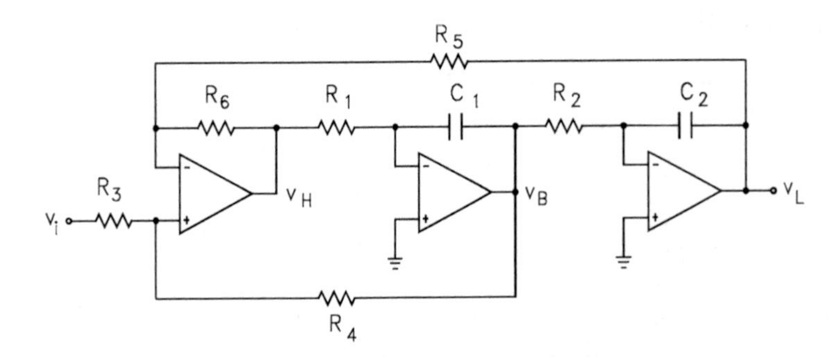

Figure 5.6: Second-order state variable filter.

$$T(s) = \frac{\omega_o^2}{s^2 + \dfrac{\omega_o}{Q} s + \omega_o^2} \tag{5.20}$$

Both the step function response and the sinusoidal response may be simulated using an electronic circuit known as a state variable filter. In this simulation voltages at the output of op-amps are analogous to the position, velocity, and acceleration of the mass.

5.2.3 State Variable Filter

Shown in Fig. 5.6 is a second-order state variable filter. There is one input, $v_i(t)$, and three outputs: $v_H(t)$, the high-pass output; $v_B(t)$, the band-pass output; and $v_L(t)$, the low-pass output. This is a filter which can be used as a low-pass filter, a band-pass filter or a high-pass filter. It can also be used to solve the second order mechanical system problem discussed above. The low-pass output is analogous to the displacement, the band-pass output is analogous to the velocity, and the high-pass output is analogous to the acceleration of the mass.

The complex transfer functions for this system are:

$$T_L(s) = \frac{V_L}{V_i} = K_L \frac{1}{1 + \dfrac{1}{Q}\left(\dfrac{s}{\omega_o}\right) + \left(\dfrac{s}{\omega_o}\right)^2} \tag{5.21}$$

$$T_H(s) = \frac{V_H}{V_i} = K_H \frac{\left(\dfrac{s}{\omega_o}\right)^2}{1 + \dfrac{1}{Q}\left(\dfrac{s}{\omega_o}\right) + \left(\dfrac{s}{\omega_o}\right)^2} \tag{5.22}$$

$$T_B(s) = \frac{V_B}{V_i} = -K_B \frac{\dfrac{1}{Q}\left(\dfrac{s}{\omega_o}\right)}{1 + \dfrac{1}{Q}\left(\dfrac{s}{\omega_o}\right) + \left(\dfrac{s}{\omega_o}\right)^2} \tag{5.23}$$

where

$$\omega_o = \sqrt{\frac{\gamma}{\tau_1 \tau_2}} \tag{5.24}$$

$$Q = \frac{1}{\alpha(1+\gamma)}\sqrt{\gamma\frac{\tau_1}{\tau_2}} \tag{5.25}$$

$$K_L = \frac{\beta}{\gamma}(1+\gamma) \tag{5.26}$$

$$K_H = \beta(1+\gamma) \tag{5.27}$$

$$K_B = \frac{\beta}{\alpha} \tag{5.28}$$

$$\gamma = \frac{R_6}{R_5} \tag{5.29}$$

$$\alpha = \frac{R_3}{R_3 + R_4} \tag{5.30}$$

$$\beta = \frac{R_4}{R_3 + R_4} \tag{5.31}$$

$$\tau_1 = R_1 C_1 \tag{5.32}$$

$$\tau_2 = R_2 C_2 \tag{5.33}$$

are the parameters of the transfer functions.

5.3 Procedure

The op-amps that will be used in this experiment are the 741 and TL071. The pinouts for these ICs are given in Exp. 3.

5.3.1 Power Supply Adjustment

Turn on the **Tektronix DMM4040** DMM (digital multimeter). Set the DMM to measure DC voltage (DCV which is the mode it boots up in). Check the power supply voltages at the binding posts to see if they are set to plus and minus 12 Volts. Turn the DMM off.

5.3.2 Integrator

Assemble the circuit shown in Fig. 5.7. Use a 741 as the op-amp. Set the function generator to produce a triangular wave with a peak value of 1 V and a frequency of 100 Hz. Turn the **Tektronix DPO3012** oscilloscope on. Connect the leads for $CH1\&2$ as shown. Press **Autoset**.

Print the display.

Change the function generator to square and repeat. Change the function generator to sine and repeat. Demonstrate the properly functioning circuit to the Laboratory Instructor.

Laboratory Instructor Verification ______________________________

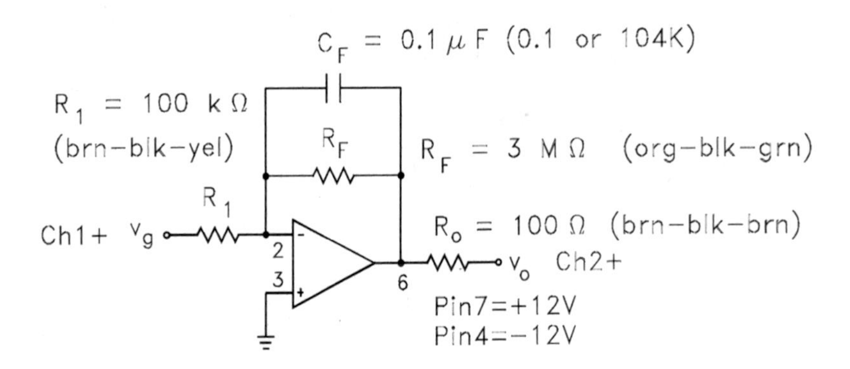

Figure 5.7: Integrator.

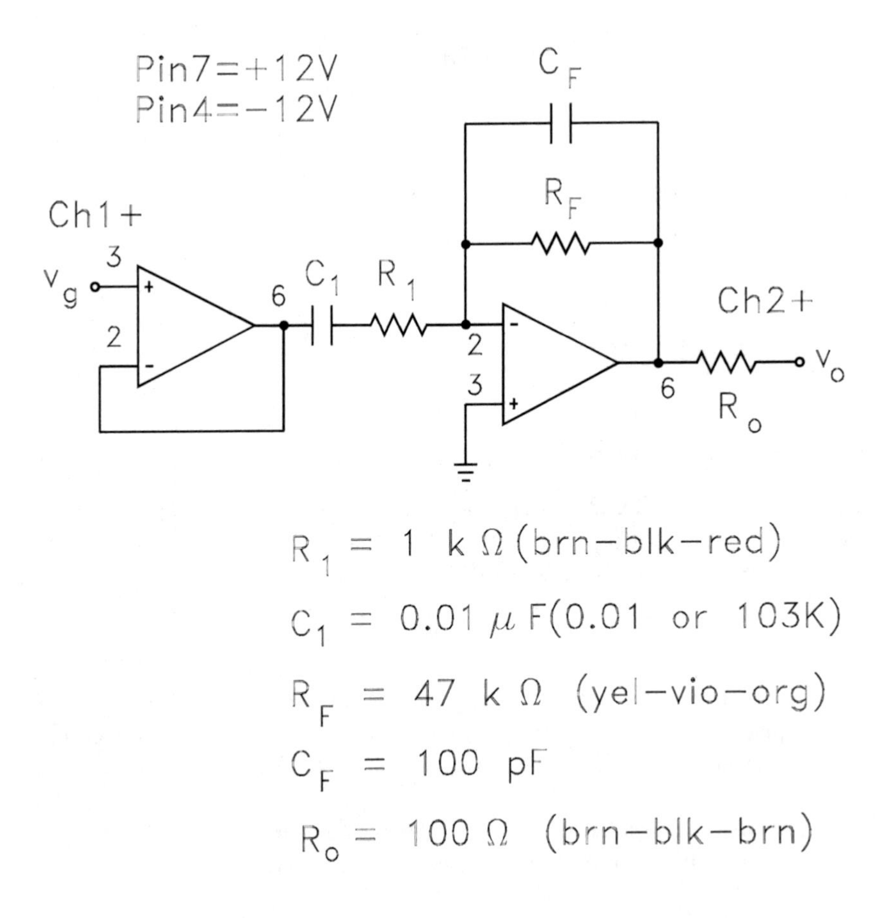

Figure 5.8: Differentiator.

5.3.3 Differentiator

Assemble the circuit shown in Fig. 5.8. Use an TL071 for the first op-amp (the one on the left which is acting as a unity gain op-amp buffer) and a 741 for the second op-amp. Set the function generator to produce a triangular wave with a peak value of 1 V and a frequency of 100 Hz. Press **Autoset** on the oscilloscope.

Print the display.

Change the function to sine and repeat. Change the function to square and repeat. Demonstrate the properly functioning circuit to the Laboratory Instructor.

Laboratory Instructor Verification ____________________

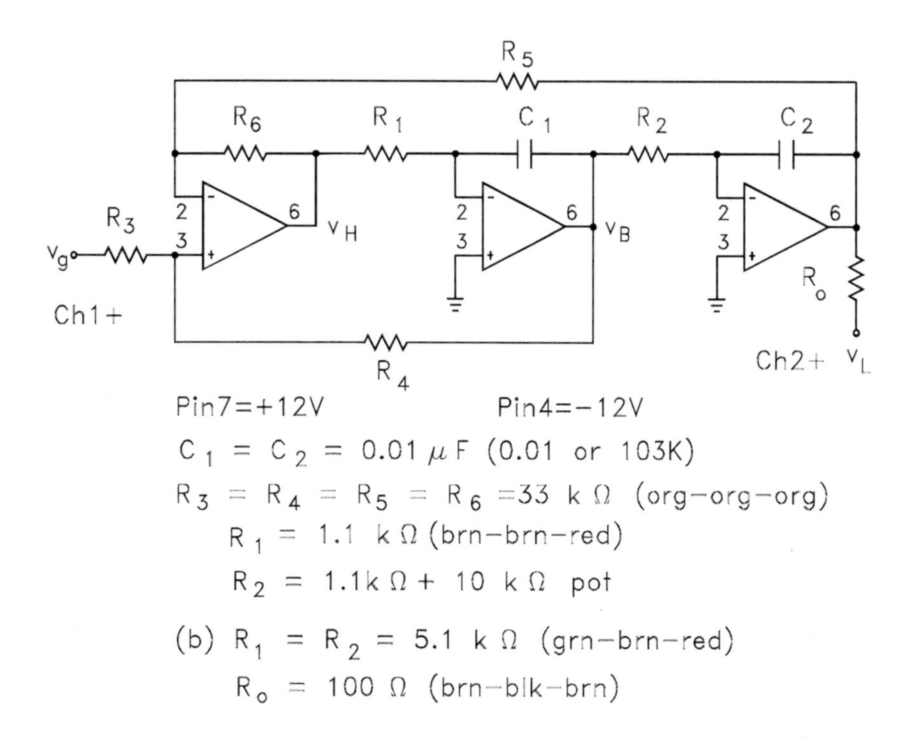

Figure 5.9: State variable filter.

5.3.4 State Variable Filter

Step Function Response

Assemble the circuit shown in Fig. 5.9. Either 741 or TL071 op-amps or any combination of the two may be used. The resistors R_1 and R_2 will be changed at several points in this procedure but the other circuit components will remain the same. Use $R_1 = 1.1\,\text{k}\Omega$ $(brn - brn - red)$ and $R_2 = 1.1\,\text{k}\Omega$ in series with a $10\,\text{k}\Omega$ pot. The $10\,\text{k}\Omega$ pot is located at the bottom of the **CADET**. Connect one side of the $1.1\,\text{k}\Omega$ resistor to the wiper of the pot. This will yield a resistor that can vary from $1.1\,\text{k}\Omega$ to $11.1\,\text{k}\Omega$.

Set the function generator to produce a 1 kHz square wave with a peak value of 1 V. Connect $CH2$ of the oscilloscope to the low-pass output. Press **Autoset** on the oscilloscope. Vary the $10\,\text{k}\Omega$ pot from one extreme to the other. Set the pot to produce critically dampening (no overshoot in the output).

Print the display or sketch the display using the graph paper provided at the end of the experiment.

Disconnect the pot from the circuit and measure the resistance value for R_2 that produced critical dampening; this is the resistance of the 10 kΩ pot from the wiper to the end plus the resistance of the fixed 1.1 kΩ resistor. The resistance is measured with the **Tektronix DMM4040** DMM; set it to the "Ω" setting and use the two inputs on the upper left. Turn off the DMM.

Laboratory Instructor Verification ______________________________

Sine Wave Response

Replace R_1 and R_2 with 5.1 kΩ resistors (*grn – brn – red*). Set the function generator to produce a sine wave with a peak value of 1 V. The gain versus frequency data will now be obtained for all three outputs. The gain is to be measured at the frequencies: 100 Hz, 300 Hz, 700 Hz, 1 kHz, 3 kHz, 7 kHz, 10 kHz, 30 kHz, 70 kHz, and 100 kHz. Move $CH2$ on the scope from the low-pass to the band-pass and then the high-pass outputs to obtain these gains.

Frequency	V_1 (*Volts*)	V_H (*Volts*)	V_B (*Volts*)	V_L (*Volts*)
100 Hz				
300 Hz				
700 Hz				
1 kHz				
3 kHz				
7 kHz				
10 kHz				
30 kHz				
70 kHz				
100 kHz				

State-Variable Filter

Demonstrate the properly functioning circuit to the Laboratory Instructor.

Laboratory Instructor Verification ______________________________

5.4 Laboratory Report

What value of resistance R_2 (Ω) produced critical dampening?

____

Does this value agree with the value predicted by the equations in the theory section? For critical dampening $\zeta = 1$ which implies the $Q = 1/2$.

____

____

Plot the gains versus frequency on the enclosed graph paper. Alternatively, the plots may be made with a spreadsheet such as Excel or K Graph.

What was the center frequency of the band-pass filter (Hz)?

Turn in all sketches and plots and the answers to all questions in the procedure as well as any questions which may have been posed by the Laboratory Instructor. Return the DMM to the Laboratory Instructor.

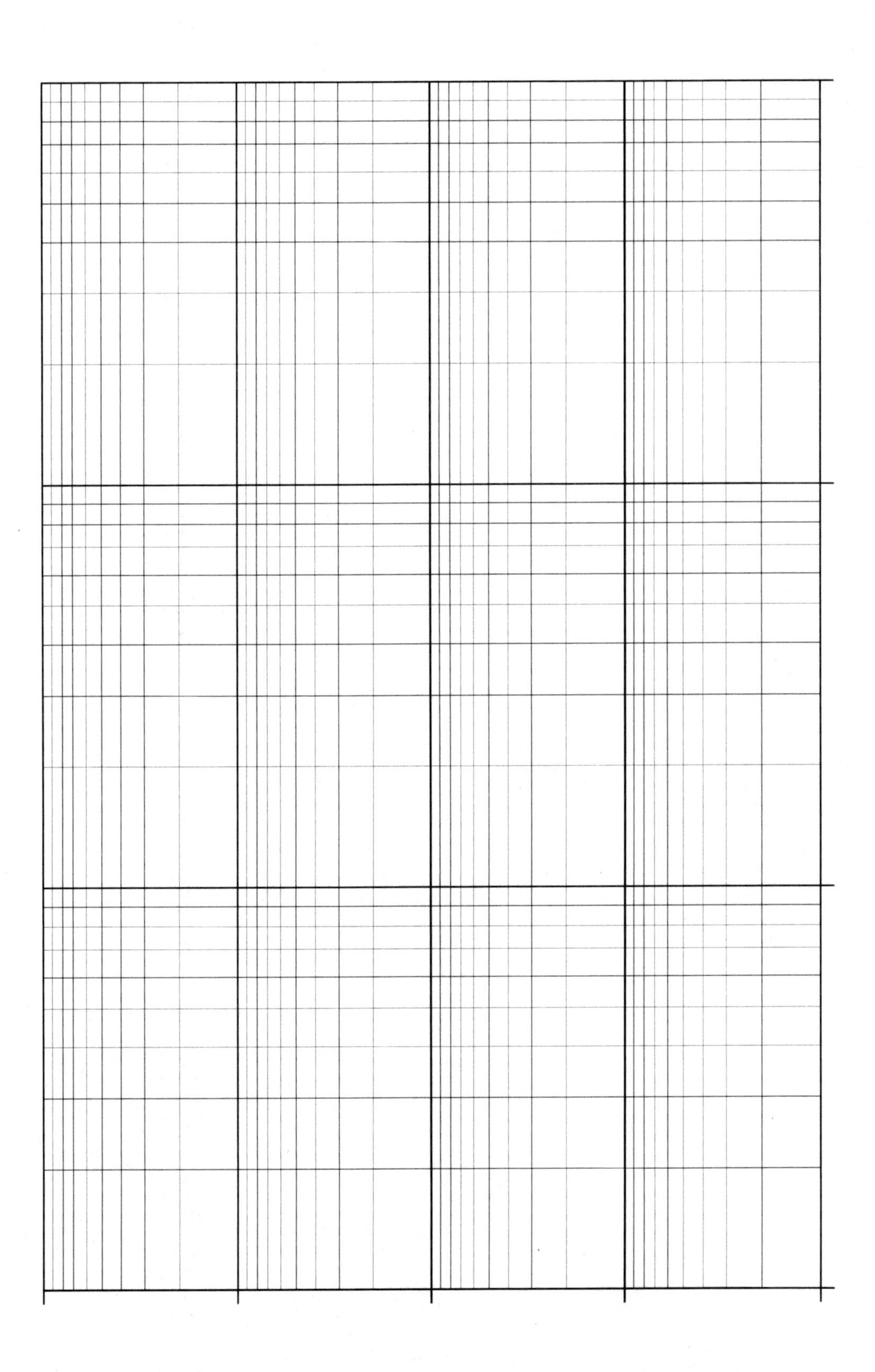

Figure 5.10:

Title:____________________

Figure No.____________________

VOLTS/DIV

Ch1 ________
Ch2 ________
Ch3 ________
Ch4 ________

TIME/DIV
MAIN

____

DELAYED

____

DELAY

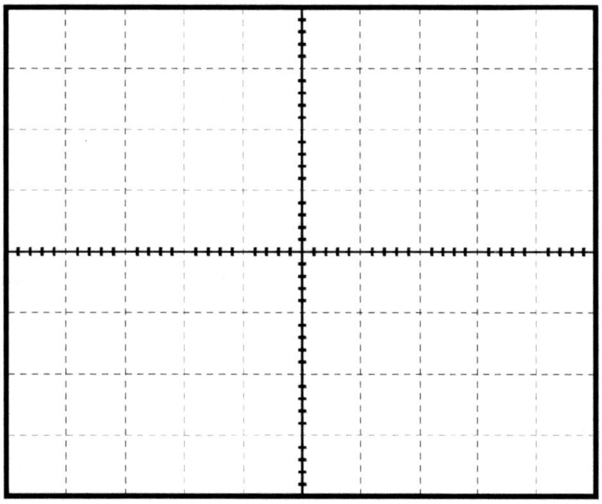

____

Coupling

AC ☐
DC ☐
Ch 2 Inv ☐
BW Limit ☐

Title:____________________

Figure No.____________________

VOLTS/DIV

Ch1 ________
Ch2 ________
Ch3 ________
Ch4 ________

TIME/DIV
MAIN

____

DELAYED

____

DELAY

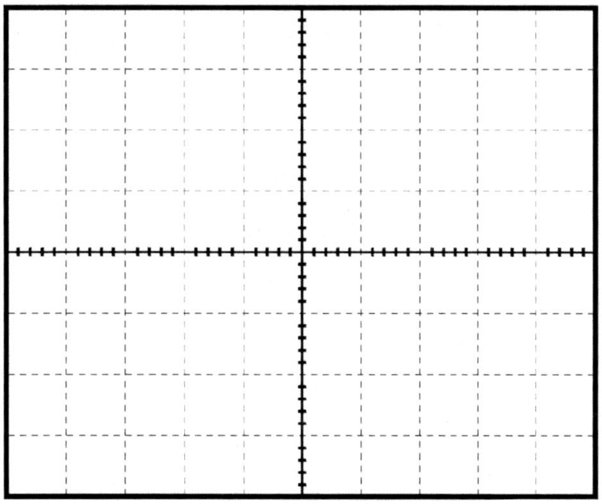

____

Coupling

AC ☐
DC ☐
Ch 2 Inv ☐
BW Limit ☐

Figure 5.11:

Chapter 6

Diodes and Op Amps

6.1 Object

This experiment examines some elementary nonlinear op amp circuits such as active rectifiers, peak detectors, and comparators which require diodes in addition to op amps.

6.2 Theory

6.2.1 Half-Wave Precision Rectifier

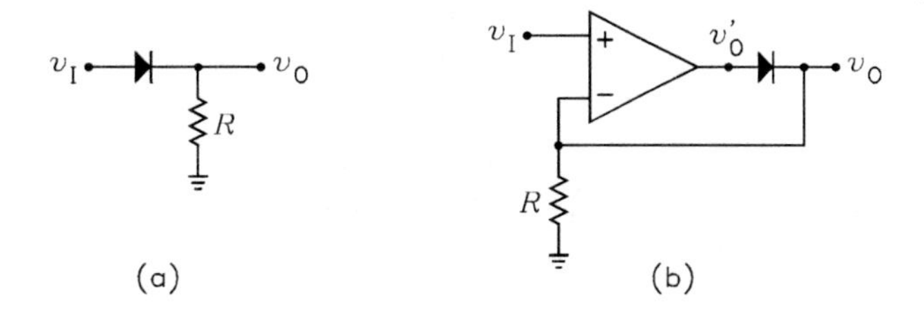

Figure 6.1: (a) Half-wave rectifier circuit. (b) Precision half-wave rectifier circuit.

Figure 6.1(a) shows the circuit diagram of a passive half-wave rectifier consisting of a diode and a load resistor. To explain the operation of the circuit, let us consider the diode to be ideal. That is, it is a short circuit if a voltage is applied to cause a current to flow in the direction of the arrow and an open circuit if a voltage is applied to cause a current to flow against the direction of the arrow. In this case, the output voltage is given by

$$v_O = v_I \quad \text{for} \quad v_I \geq 0 \tag{6.1}$$

$$= 0 \quad \text{for} \quad v_I < 0 \tag{6.2}$$

The circuit is called a half-wave rectifier because it passes the signal only on its positive cycle. If the diode is reversed, it pass the signal on its negative cycle. The voltage across a physical diode is not zero when it is forward biased but varies logarithmically with the current. This degrades the performance of the rectifier unless the applied voltage is large compared to the voltage drop across the diode.

Many of the limitations of physical diodes in rectifier circuits can be overcome with op amps. Such a circuit is shown in Figure 6.1(b). For $v_I > 0$, the op amp output voltage v'_O is positive. This forward biases the diode and causes v_O to go positive. Because the op amp has negative feedback, the difference voltage between its two inputs is forced to zero, making $v_O = v_I$. For $v_I < 0$, the voltage v'_O is negative. This reverse biases the diode. The current flow through it is essentially zero, making $v_O = 0$. It follows that the output voltage v_O is given by Eq. (6.1).

The circuit of Figure 6.1(b) has the disadvantage that the op-amp loses feedback when $v_I < 0$. When this happens, the op amp gain increases to its open-loop value and v'_O falls to the negative saturation voltage $-V_{SAT}$. When v_I again goes positive, v'_O must increase from $-V_{SAT}$ to a positive voltage before the diode becomes forward biased. Because the op amp slew rate is not infinite, v'_O cannot change instantaneously. This causes a time delay before the diode becomes forward biased, thus degrading the operation of the circuit. This degradation becomes worse as the frequency of the input signal is increased.

Figure 6.2 shows an improved circuit for which the op amp does not saturate. For $v_I > 0$, v'_O goes negative. This causes D_1 to be reverse biased and D_2 to be forward biased. The negative feedback through R_F forces the voltage at the inverting input to be zero. To solve for v_O, we can write $i_1 + i_F = 0$, $i_1 = v_I/R_1$, and $i_F = v_O/R_F$. Solution of these equations yields $v_O = (-R_F/R_1)\, v_I$. For $v_I < 0$, v'_O goes positive. This causes D_1 to be forward biased and D_2 to be reverse biased, making $i_F = 0$. The op amp has negative feedback through D_1 which forces the voltage at the inverting input to be zero. It follows that $v_O = 0$. Thus the circuit operates as a precision half-wave rectifier. The voltage gain is $-R_F/R_1$. The negative sign means that the gain is inverting. Because the op amp does not saturate when $v_O = 0$, the circuit does not have the limitations of the one in Figure 6.1(b). If the direction of each diode is reversed, $v_O = 0$ for $v_I > 0$ and $v_O = (-R_F/R_1)\, v_I$ for $v_I < 0$. Again, the circuit has an inverting voltage gain.

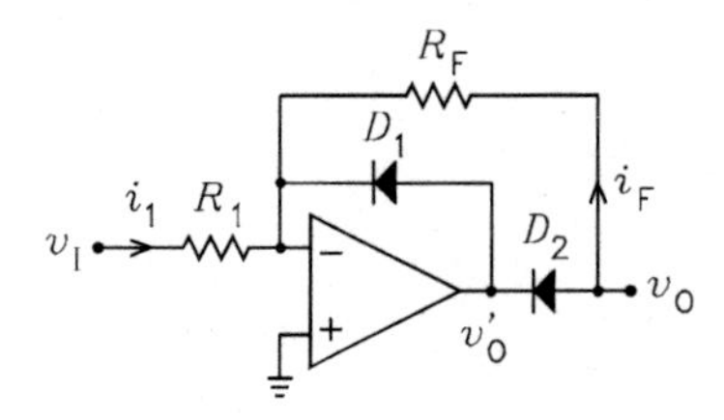

Figure 6.2: Improved precision half-wave rectifier.

6.2.2 Full-Wave Precision Rectifiers

In many applications, a full-wave rectifier is preferred over a half-wave rectifier. The output voltage from a full-wave rectifier is given by

$$v_O = +k\,|v_I| \qquad \text{or} \qquad v_O = -k\,|v_I| \tag{6.3}$$

where k is the gain. The plus sign is used for a non-inverting gain and the minus sign is used for an inverting gain. This section describes four precision full-wave rectifier circuits.

Figure 6.3 shows the circuit diagram of a commonly used precision full-wave rectifier circuit. It consists of the half-wave circuit of Figure 6.2 followed by an inverting summer circuit. The output voltage is given by $v_O = (-R_F/R_2)\, v_I + (-2R_F/R_2)\, v_{O1}$. For $v_I > 0$, $v_{O1} = -v_I$ so that $v_O = (-R_F/R_2)\, v_I + (2R_F/R_2)\, v_I = (R_F/R_2)\, v_I$. For $v_I < 0$, $v_{O1} = 0$ so that $v_O = (-R_F/R_2)\, v_I$. These results can be combined to obtain

$$v_O = \frac{R_F}{R_2} \times |v_I| \tag{6.4}$$

If the direction of each diode is reversed, the output is multiplied by -1.

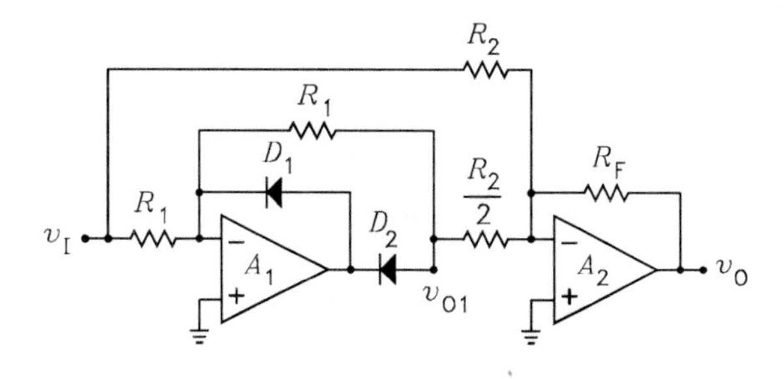

Figure 6.3: Precision full-wave rectifier.

6.2.3 Peak Detector Circuit

Figure 6.4 shows the circuit diagram of a peak detector. To explain its operation, let us assume that R_1 is removed from the circuit. Let the initial voltage on the capacitor be zero. If v_I goes positive, v_{O1} will go positive. This forward biases D_1 and reverse biases D_2. The capacitor charges positively, thus causing v_O to increase. Because there is no current through R_2, the voltage at the inverting input of A_1 is v_O. The negative feedback around A_1 causes the voltage difference between its inputs to be zero. Thus it follows that $v_O = v_I$.

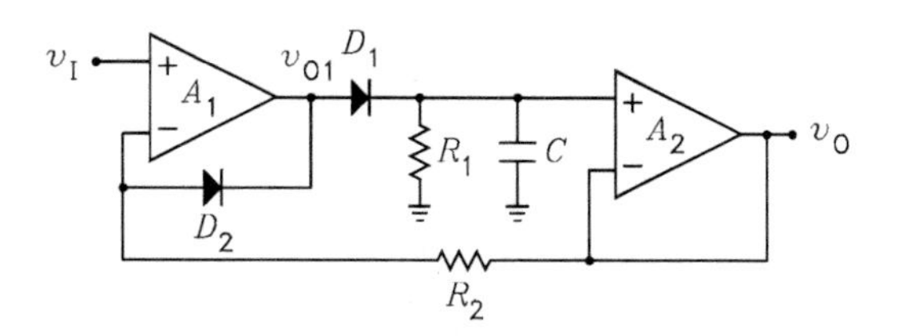

Figure 6.4: Peak detector circuit.

Now let v_I decrease from its previous positive peak value. This causes v_{O1} to decrease, causing D_1 to cut off. When this happens, A_1 loses feedback and its output voltage decreases rapidly, causing D_2 to become forward biased. R_2 limits the current in D_2 when this occurs. The capacitor voltage cannot change because there is no discharge path with R_1 removed from the circuit. Therefore, v_O remains at the previous peak value of v_I. If v_I increases again to a value greater than v_O, D_1 will be forward biased and D_2 will be reverse biased, causing the capacitor to charge to a higher positive voltage. Thus the circuit acts as a peak detector which holds the highest positive peak voltage applied to its input. When D_1 is reverse biased, its reverse saturation current and the input bias current of A_2 can cause the capacitor voltage to change. Therefore, low leakage current diodes and op amps with a JFET input stage, i.e. a bifet op amp, should be used. If the directions of the diodes are reversed, the circuit detects the negative peaks of the input signal.

With R_1 in the circuit, the capacitor discharges with a time constant $\tau = R_1 C$ when D_1 is reverse biased, causing v_O to decrease toward zero. However, D_1 becomes forward biased again when $v_O = v_I$, thus causing v_O to follow the positive peaks in the input as opposed to holding the peaks. Very large discharge time constants require large values for R_1, for C, or for both. Large value capacitors should be avoided in this circuit because A_1 can be driven into current limiting when the capacitor is charged. This can cause the circuit not to detect the full peak value of the input. All physical capacitors exhibit a leakage resistance R_P in parallel with the capacitor. This resistor appears in parallel with R_1 in the circuit. If R_P cannot be neglected, the effective discharge time constant is $\tau = (R_1 \| R_P)\, C$.

6.2.4 Comparators

An op amp that is operated without feedback or with positive feedback is called a comparator. Figure 6.4 shows the circuit diagram of an op amp operated as a non-inverting comparator. For $v_I > V_{REF}$, the output voltage is $+V_{SAT}$. For $v_I < V_{REF}$, the output voltage is $v_O = -V_{SAT}$. Thus we can write the general relation

$$v_O = V_{SAT}\text{sgn}\left(v_I - V_{REF}\right) \tag{6.5}$$

where sgn(x) is the signum function defined by

$$\begin{aligned} \text{sgn}\,(x) &= +1 \quad \text{for} \quad x > 0 \\ &= -1 \quad \text{for} \quad x < 0 \end{aligned} \tag{6.6}$$

If the inputs to the op amp are reversed, it becomes an inverting comparator and the output voltage is multiplied by -1.

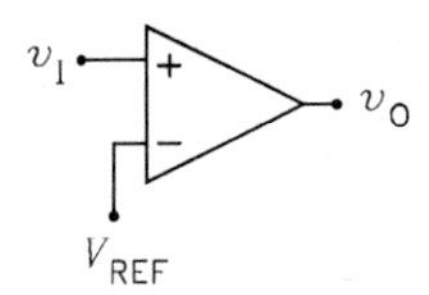

Figure 6.5: Op amp used as a non-inverting comparator.

Several comparators can be used to realize a bar level indicator. When such an indicator is used to display the level of an audio signal, it is sometimes called an active VU meter, where VU stands for volume units. Figure 6.6 shows the circuit diagram of a bar level indicator circuit having 10 levels. The circuit is realized with the LM3914 Dot/Bar Display Driver and the LTA-1000 Bar Graph Array. The LM3914 is an 18-pin DIP (dual-in-line package) integrated circuit and the LTA-1000 is a 20-pin DIP that contains 10 LEDs (light emitting diode).

The LM3914 contains an internal voltage reference of 1.25 V that connects between pins 7 and 8. This source, in conjunction with resistors R_1 and R_2, is used to set the reference voltage V_{REF} that is applied to pin 6. Because the voltage across R_1 is 1.25 V, the current through R_1 is $1.25/R_1$. This current plus the current I_{ADJ} flows through R_2 to circuit ground. Thus the voltage V_{REF} across R_1 and R_2 is given by

$$V_{REF} = 1.25 + \left(\frac{1.25}{R_1} + I_{ADJ}\right) R_2 = 1.25\left(1 + \frac{R_2}{R_1}\right) + I_{ADJ}R_2 \tag{6.7}$$

The current I_{ADJ} is specified to be less than 120 μA which is small enough to usually neglect in the equation. The signal input is connected to pin 5 which connects to a voltage follower buffer. The output of the buffer connects to the inverting input of each of the 10 voltage comparators.

The reference voltage for each comparator is derived from a voltage divider consisting of ten 1 kΩ resistors connected in series. For a zero signal input, the output voltage of each comparator is $+V_{SAT}$. This reverse biases all of the LEDs so that none are on. If the buffer output voltage increases to a level that is greater than the reference voltage of any comparator, the output voltage from that comparator falls to zero, thus forward biasing the LED connected to that comparator. This causes the LED to emit light. Not shown in the figure is a resistor in series with each LED that limits the current when the comparator output voltage falls to zero.

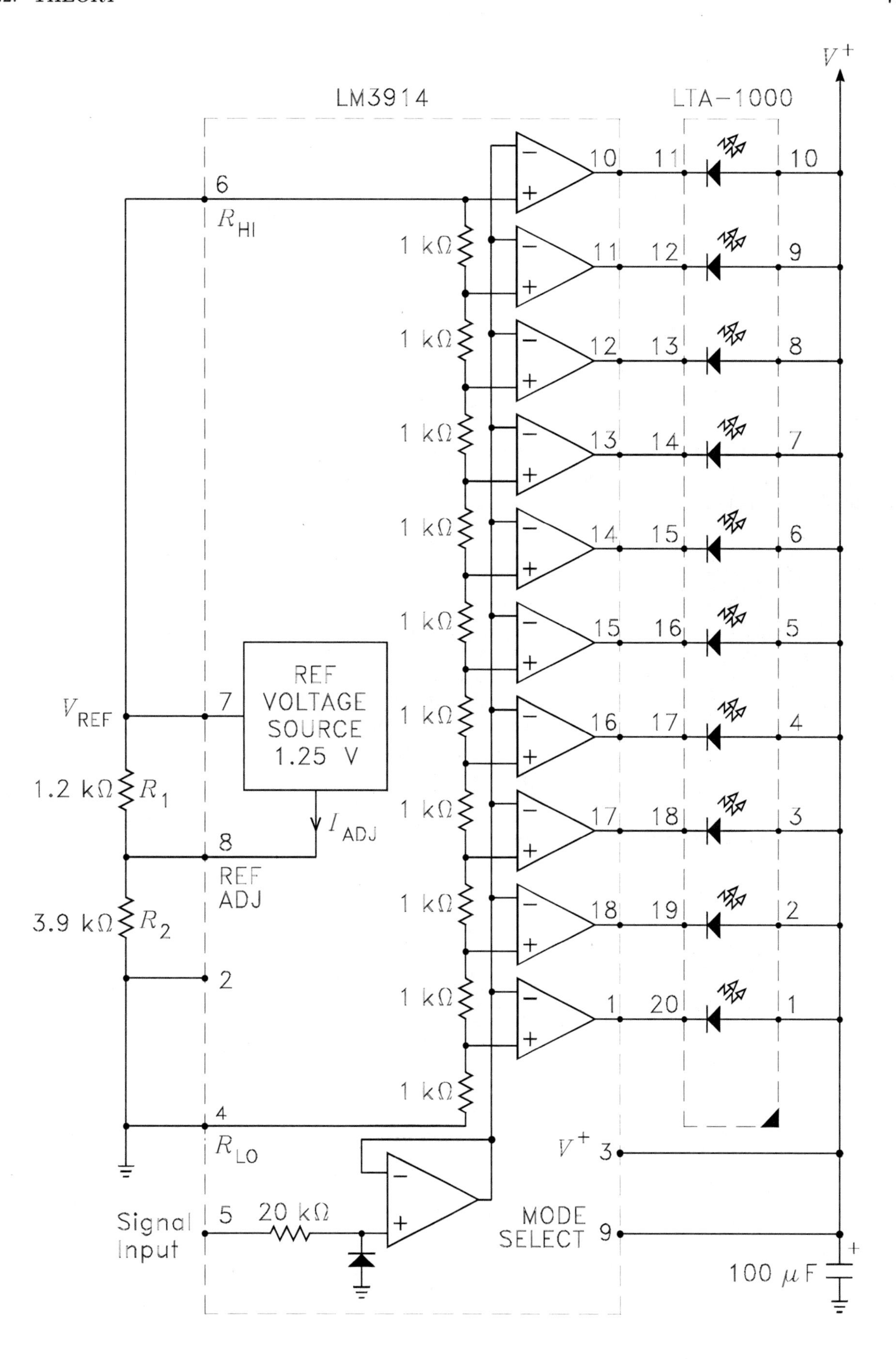

Figure 6.6: Bar-graph level indicator.

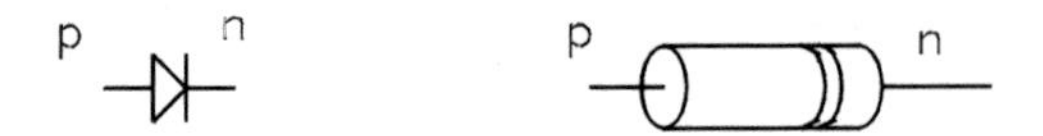

Figure 6.7: Diode circuit symbol and lead arrangement.

6.3 Procedure

6.3.1 Rails

Establish rails or buses for +12 V, −12 V, and ground. Use the long vertical buses parallel to the IC trough. Remember to put jumpers at the midpoint of the vertical buses so that they will extend from the top to the bottom of the breadboard.

6.3.2 Diodes

The type of diode used in this experiment is the 1N4148 diode. This is known as a signal diode. It does not emit visible light and is used in low power applications. The circuit symbol and pin or lead connections are shown in Fig 6.7. The triangle in the circuit symbol is the p side and the vertical line is the n. On the physical cylindrical package the n side is indicated with a band.

Turn on the **Tektronix DMM4040** DMM and wait for it to boot. Press the symbol for a diode. For each of the two diodes used in this experiment, connect the HI lead of the DMM to the p side and the LO lead to the n side. A low resistance should be obtained. Reverse the leads and a high resistance should be obtained. If this is not the case the diode is probably defective.

6.3.3 Passive Half-Wave Rectifier

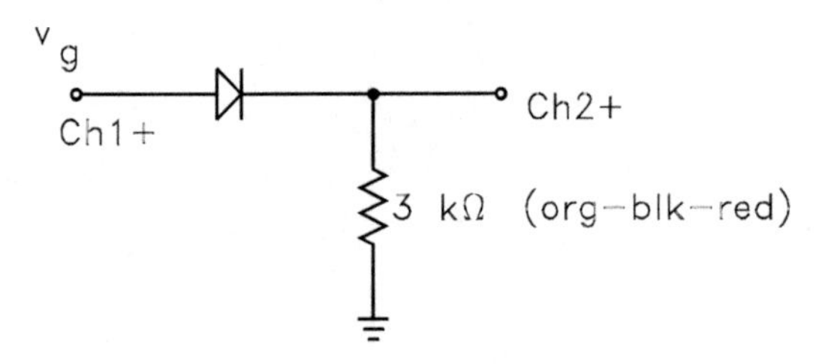

Figure 6.8: Passive rectifier.

Assemble the circuit shown in Figure 6.8 with the power on the **CADET** turned off. Use a 1N4148 or equivalent for the diode. Turn on the **Tektronix DPO3012** oscilloscope and wait for it to boot. Turn on the **CADET**. Set the function generator to produce a sine wave with a frequency of 100 Hz, and a peak-to-peak value of 2 V. Press *AUTOSET* on the oscilloscope and adjust the controls on the **CADET** until the input signal is correct.

What is the peak value of the output?____________________

What is the duty cycle?____________________

The duty cycle is the percentage of one period or one cycle for which the output is nonzero. Use the time cursors to measure the time interval for which the output is nonzero and then divide by the period (reciprocal of the frequency) and multiply by 100 to express it as a percentage. To activate the cursors press *CURSOR*, *V Bars*, *Bring both Cursors on Screen*. The *SELECT* knob toggles which cursor is active and its position is then controlled with the rotary knob. Print or sketch the display.

Set the oscilloscope for XY mode to plot the output versus the input; set the sensitivity for the horizontal and vertical axes so that the transfer characteristic can be adequately displayed (both to 0.2 V per division). (To set the XY mode press the *DISPLAY*, *XY Display*, *Triggered XY*.) Print or sketch the display.

Laboratory Instructor Verification ______________________________

6.3.4 Precision Half-Wave Rectifier

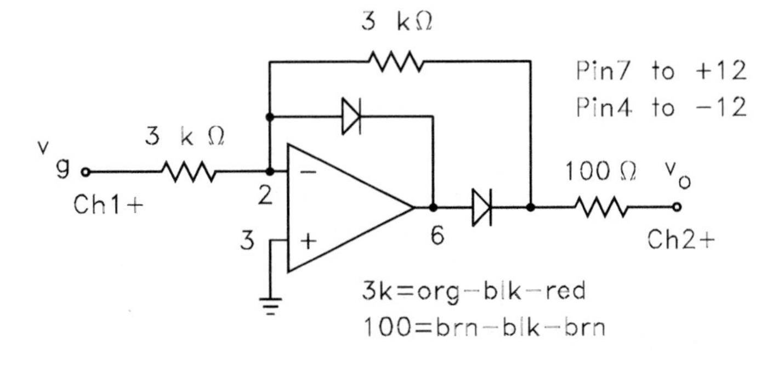

Figure 6.9: Precision half-wave rectifier.

Assemble the circuit shown in Figure 6.9 with the power on the **CADET** turned off. Use either a 741 or TL071 for the op amp and 1N4148s or equivalent for the diodes. Turn on the **CADET**. Set the function generator to produce a sine wave with a frequency of 100 Hz, and a peak-to-peak value of 2 V. Press *AUTOSET* on the oscilloscope and adjust the controls on the **CADET** until the input signal is correct.

What is the peak value of the output?____________________

What is the duty cycle?________________________

The duty cycle is the percentage of one period or one cycle for which the output is nonzero. Use the time cursors to measure the time interval for which the output is nonzero and then divide by the period (reciprocal of the frequency) and multiply by 100 to express it as a percentage. Print or sketch the display.

Set the oscilloscope for XY mode to plot the output versus the input; set the sensitivity for the horizontal and vertical axes so that the transfer characteristic can be adequately displayed (both to 0.2 V per division). Print or sketch the display.

Laboratory Instructor Verification ____

6.3.5 Full-Wave Rectifier

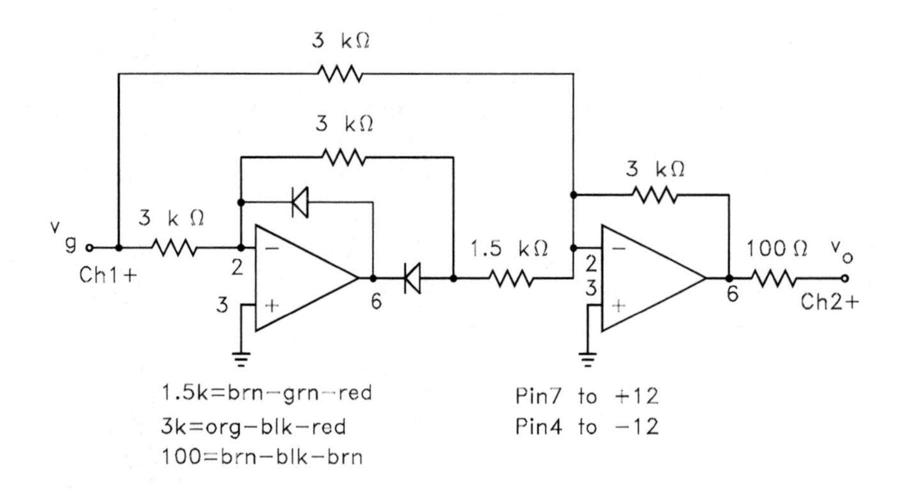

Figure 6.10: Precision full-wave rectifier.

Assemble the circuit shown in Figure 6.10 with the power on the **CADET** turned off. Use either 741s or TL071s for the op amps and 1N4148s or equivalent for the diodes. (Note that the direction of the diodes has been reversed from the previous circuit.) Turn on the **CADET**. Set the function generator to produce a sine wave with a frequency of 100 Hz, and a peak-to-peak value of 2 V. Press Auto-Scale on the oscilloscope and adjust the controls on the **CADET** until the input signal is correct.

What is the peak value of the output?________________________

What is the duty cycle?________________________

The duty cycle is the percentage of one period or one cycle for which the output is nonzero. Use the time cursors to measure the time interval for which the output is nonzero and then divide by the period (reciprocal of the frequency) and multiply by 100 to express it as a percentage. Print or sketch the display.

Set the oscilloscope for XY mode to plot the output versus the input; set the sensitivity for the horizontal and vertical axes so that the transfer characteristic can be adequately displayed (both to 0.2 V per division). Print or sketch the display.

Laboratory Instructor Verification ____

6.3.6 Peak Hold Circuit

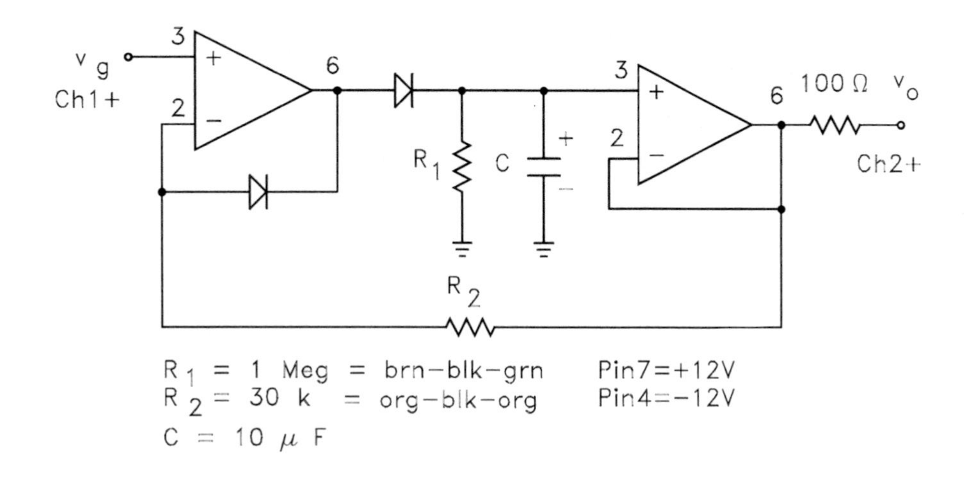

Figure 6.11: Peak hold circuit.

Assemble the circuit shown in Figure 6.11 with the power on the **CADET** turned off. Use either 741s orTL071s for the op amps and 1N4148s or equivalent for the diodes. Use a value of 30 kΩ for R_2, $C = 10\ \mu$F, and $R_1 = 1$ MΩ. Note that C is an electrolytic capacitor and must be inserted with the correct polarity. Turn the power on the **CADET** on. Set the function generator to produce a sine wave with a frequency of 100 Hz, and a peak-to-peak value of 4 V. Connect the function generator to the input of the circuit and to $CH1$of the oscilloscope. Connect the output of the circuit to $CH2$ of the oscilloscope. Set the sensitivity to 1 V/div for each channel and use the vertical position controls so that the center horizontal grid line is zero volts for each channel. The zero volt line is indicated to the left of the display for each channel with an icon with the color coded channel number. Set the time base to display several cycles of function generator voltage. The output of the circuit should be a straight line touching the positive peak of the input sine wave. Sketch or print the display. Note the effect of varying the input amplitude up and down on the function generator. Why does it go up faster than down?

Laboratory Instructor Verification ____

6.3.7 Bar Graph Array

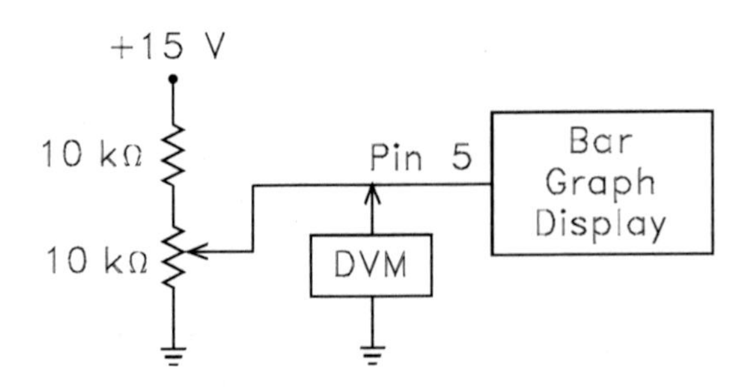

Figure 6.12: Bar graph circuit.

Assemble the circuit shown in Figure 6.12 with the power on the **CADET** turned off. The portion of this figure which is labeled as "Bar Graph Display" is Figure 6.6. The LM3914 is a standard 18 pin DIP IC. Note that the pin locations shown in Figure 6.6 are grouped by function rather than physical location. The pins locations shown for the LTA-1000 DIP Bar Graph Array are grouped by their physical location. Three corners of the LTA-1000 are square while the remaining one is beveled or rounded and it is this corner that indicates pin 1. The portion of Figure 6.6 inside the dotted lines are internal to the devices while the portion outside the dotted lines are external components or connection which must be made. Note that the anode or p side of each LED must be directly connected positive binding post of the dc power supply; i.e. the power supply decoupling network must be modified because of the excessive current that the LEDs will draw when lighted. Note that the 100 μF capacitor is an electrolytic capacitor which must be inserted with the correct polarity. The terminal indicated as V^+ is the positive power supply rail, +12 V. No connection is made to the negative power supply.

The 10 kΩ pot is located at the bottom of the CADET. The arrow is known as the wiper. One side of the pot is connected to the positive power supply rail, +12 V, and the other to ground. The wiper of the pot is connected to pin 5 of the LM3914. The DVM (Digital Voltmeter) is the **Agilent 34401A** DMM configured to measure dc voltage. Connect the HI lead to pin 5 of the LM3914 and the LO lead to ground.

Turn the **CADET** on. Vary the potentiometer and measure and record the input voltage at which each of the 10 segments of the bar graph array lights. In the below table segment number 1 is the one at the bottom of the bar graph display adjacent to pin 1.

Segment Number	DC Voltage
10	
9	
8	
7	
6	
5	
4	
3	
2	
1	

Laboratory Instructor Verification ____________________

6.4 Laboratory Report

Turn in all sketches taken or printouts made and a tabular summary of all data taken. Answer any questions posed in the procedure or by the laboratory instructor.

Title:________________________

Figure No.________________________

VOLTS/DIV

Ch1 __________
Ch2 __________
Ch3 __________
Ch4 __________

TIME/DIV
MAIN

____

DELAYED

____

DELAY

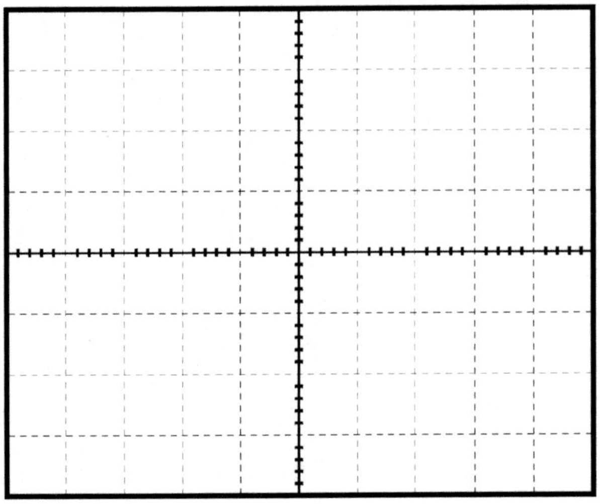

____

Coupling
AC ☐
DC ☐
Ch 2 Inv ☐
BW Limit ☐

Title:________________________

Figure No.________________________

VOLTS/DIV

Ch1 __________
Ch2 __________
Ch3 __________
Ch4 __________

TIME/DIV
MAIN

____

DELAYED

____

DELAY

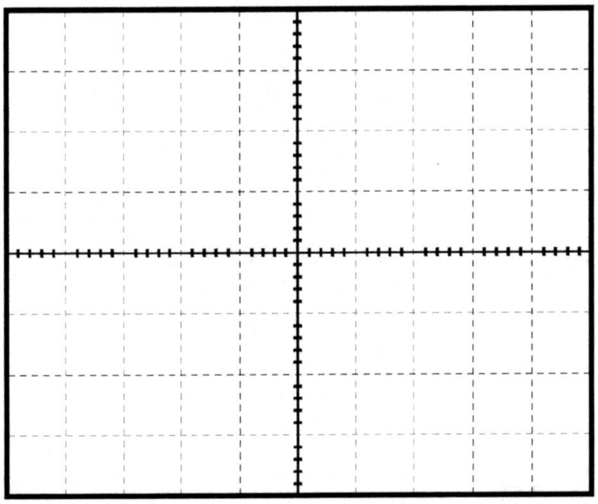

____

Coupling
AC ☐
DC ☐
Ch 2 Inv ☐
BW Limit ☐

Figure 6.13:

Chapter 7

Introduction to Combinational Logic

7.1 Objective

The objective of this experiment is to experimentally examine some elementary combinational logic gates and some common applications of these devices. The gates that will be examined are the *INVERTER*, *AND*, *NAND*, *OR*, *NOR*, and *EXOR*. The applications that will be investigated are the half and full adders, the Gray code encoder and decoder, and the implementation of arbitrary logic equations.

7.2 Theory

7.2.1 Introduction

Technological historians generally bestow the honor of having conceived the idea of the first general-purpose programmable computer to the English mathematician Charles Babbage (1791-1871). Babbage's computer was known as the Analytic Machine (1834) and had a "store" which could hold 100 48 digit numbers and a "mill" which processed numbers. Numbers were held in the store until needed by the mill and were then recalled and sent back to the store or printed. This machine was to be programmed by punched cards.

Unfortunately the technology simply didn't exist to construct Babbage's machine. It was to be steam powered, the size of a railroad locomotive, and was to consist of a intricate mass of intermeshing metal clockwork. The most skilled mechanical and civil engineers of that or any other day simply could not develop a sophisticated mechanical computer. All attempts to develope a nonelectronic computer resulted in a machine with an effectiveness ranging from nonfunctional to anile. The development of a computer from idea to reality had to await the development of electrical engineering and, particularly, electronics as mature disciplines.

Babbage was an eccentric and reclusive individual. Although he was a professor of mathematics at Cambridge University (holding the same chair that was once held by Sir Isaac Newton) he never delivered a single lecture or even visited the university. Babbage's work would probably have been lost to history had it not been extensively documented by his friend and companion Countess Augusta Ada Lovelace (1815-1852). She had considerable mathematical abilities and is credited with having written the world's first computer program which, in a certain sense, makes her history's first truly calculating female. Countess Lovelace's other claim to fame is that she was the only legitimate child of the celebrated English poet and bon vivant Lord Byron (1788-1824).

A mathematical theory of logic was developed by George Boole in 1854. In this branch of mathematics variables have only two values corresponding to true and false. This is the mathematics that is used in modern day computers and digital logic. Appropriately, it is known as Boolean algebra. Boole was a poor individual who couldn't afford to attend school. He taught himself Latin, Greek, French, German, Italian, and much of the mathematics that was known at that time. He became a university professor at Queen's

College in Ireland even though he didn't even have the equivalent of a grade school education.

In 1936 Claude Shannon demonstrated that electronic circuits could be used to implement Boolean logic. Shannon is a revered figure in electrical engineering because of his pioneering work in the development of the mathematical theory of communications and, in particular, founding a discipline known as information theory.

The first devices that could be called computers were developed in World War II when the need for them became urgent. The Mark I computer was developed by Howard Aiken using Babbage's general concepts in 1943. This computer used electromechanical relays, was 51 feet long and 8 feet high and contained 750,000 parts which were interconnected with 500 miles of wire. This computer was used for ballistic studies or fire control by the Navy. Aiken was a Harvard mathematician who developed this computer while working for IBM. Aiken and the president of IBM, Thomas Watson, developed such a hatred for each other that they refused to be in the same room with each other and today Aiken's is given a low a profile as possible in the history of IBM. Much of IBM's prominence in computers can be attributed to Watson's desire to obliterate the memory of Aiken. Such antipathy was common among the developers of the computer due to vanity and the desire to protect patent rights.

The architecture used by modern computers was stated by John von Neumann in 1945. It was he who refined Babbage's concepts by stating that a computer was to perform one operation at a time and have 5 key components: a central arithmetic logic unit, a central control unit, a memory, an input unit, and an output unit. He was a stereotypical absent minded professor who became so engrossed in mathematical problems that he was solving that, when on a trip, he would often have to phone home several times to find out why he had taken the trip. Although he took great pride in being able to quote verbatim from texts that he had read decades earlier, he couldn't remember where the drinking glasses were kept at his house. Much of modern research in computers involves trying to find architectures other than that proposed by von Neumann.

The first large scale electronic digital computer was ENIAC (Electronic Numerical Integrator and Computer) which debuted in 1946. It was 18 feet high and 80 feet long contained and 17,468 vacuum tubes. Vacuum tubes were early amplifying and switching elements which were relatively large and consumed a considerable amount of electrical power. Despite a massive cooling system the temperature of the room containing ENIAC often reached $120°F$. The failure rate of these tubes was relatively high which meant that maintaining such a computer was no trivial task.

The first important development in the miniaturization of the computer was the development of the transistor in 1947 by John Bardeen, William Shockley, and Walter Brattain. This was a solid state device made out of devices known as semiconductors which was much smaller than a vacuum tube, consumed considerably less power, and was much more reliable. Transistors quickly replaced vacuum tubes as the fundamental elements in computers in the 1950s.

The first integrated circuit was developed by Jack St. Clair Kirby in 1961 at Texas Instruments. This was an entire electronic circuit on a single chip of semiconductor. The main reason that Kirby was working at TI was that it was the only company that offered him a job when he graduated from the University of Illinois in 1958.

The first microprocessor was developed by Ted Hoff of Intel in 1971. These "computers on a chip" have found applications from automatic coffee makers, autofocus cameras, digital watches, blood pressure monitors, computerized cars, etc. and, of course, computers.

In 1977 three mass-market personal computers were introduced: the Apple II, the Radio Shack TR-80, and the Commodore PET. The Apple I sold only 200 units; its successor the Apple II made the owners of the company, Steven Jobs and Stephen Wozniak, millionaires. Radio Shack decided for an initial run of 3,500 TR-80s–one for each of their stores in the USA–so that if it didn't sell they could use it for accounting and inventory; it sold.

In 1981 IBM entered the personal computer market with its IBM PC. Prior to the PC, computers were almost the mortmain of IBM. Since IBM was, by far, the most prestigious computer manufacturer in the world, this gave enhanced legitimacy to the personal computer. This also gave birth to a new industry known as the IBM clone.

In the 1950s the first discrete transistors cost $ 8.00 each. Today a far superior transistor can be purchased for about 15 cents and an integrated circuit with hundreds of transistors cost less than a dollar. This continued miniaturization of the size, power requirements, and price of electronic components has made it possible for students to own pocket calculators that have more computing ability and data storage capacity than many of the main frame computers that were in use in the 1960s and to own digital watches that are considerably more accurate than the chronometers that were used by Captain Cook to circumnavigate the globe in the Eighteenth Century.

It is difficult to predict what the future may bring. In 1950, technological forecasters were predicting that by the end of the Twentieth Century every American home would have a small nuclear reactor in the basement and a solar powered car. Few predicted that every home would have a computer because a computer was the size of a house and, besides, what would anyone other than a crazed scientist want with a computer in their home? It is predicted that future computers may use optical signal processing or even organic material. But for now, and the foreseeable future, all computers and other digital systems are inherently electrical in nature, and, therefore, it behooves all engineers, whatever their discipline, to develop a familiarity with some of the basic elements of this technology.

The above is not intended to be a complete or even representative history of the development of computers and digital electronics. Such a subject would require a complete text.

This experiment and the next three will not involve anything as exotic or complicated as constructing a computer or other sophisticated digital system due to the limitation of scope, time, and budget. Nor will it involve using VLSI (very large scale integration) fabrication that is used in microprocessors. Instead, simple experiments using SSI (small scale integration) and MSI (medium scale integration) integrated circuits will be performed. These experiments, however, involve the same basic operations that are found in more sophisticated systems and, therefore, have considerable pedagogical value.

7.2.2 Discrete Data and Number Systems

Data of interest to engineers and scientists often occurs naturally in discrete or digital form. Some common examples are whether a switch is open or closed, a seat belt is buckled, a can is on a conveyor belt, a missile is to be launched, a key on a keyboard has been pressed, an elevator door is closed, etc. Any data which can be placed into a one to one correspondence with integers in a number system is discrete data.

Oftentimes analog data can be processed and/or stored more efficiently in digital form. Circuits known as analog to digital converters are used to convert the analog data to digital form and after it is processed in digital form, if required, it is then converted back into analog form using a circuit known as a digital to analog converter. For instance, music compact discs systems sample the analog music signal from a microphone every 23 μsec, quantize and convert the analog voltage into a 16 bit binary word, and encode and store this data on a 5 inch metal disc which stores the binary digits in the form of pits and spaces using a recording laser; the user then retrieves this music from the stored digital form by illuminating the spinning disk with another laser to obtain the binary digits stored on the disk and then passing the stored 16 bit digital words through a digital to analog converter which is then filtered and amplified before being applied to a loudspeaker.

Digital electronics uses voltage levels in electronic circuits to represent discrete data which represents either the state of a system or integers in a number system. Conceptually, there is no reason why electronic circuits with ten voltage levels could not be used to represent the familiar decimal number system. However, modern electronic technology limits digital electronics to the binary number system since it is easiest for such circuits to be in one of two states corresponding to the "1" and "0" of the binary number system. Moreover, the binary number system is the natural number system for decisions or events which can be formulated as either true or false propositions such as whether a switch is open or closed. Therefore, on the circuit level the binary number system is the one universally employed.

7.2.3 Number Systems

Number systems represent numbers as sums of weighted powers of the base. Namely, any number can be represented as

$$(a_n a_{n-1} \cdots a_1 a_0 . a_{-1} a_{-2} \cdots)_b =$$

$$a_n b^n + a_{n-1} b^{n-1} + \cdots + a_1 b^1 + a_o b^o .\ a_{-1} b^{-1} + a_{-2} b^{-2} + \cdots \tag{7.1}$$

where b is the base or radix of the number system and the weights a_i have values ranging from 0 to $b - 1$ which are determined by their distance from the decimal point. For instance in the familiar decimal number system the number

$$1943.5 = 1 \times 10^3 + 9 \times 10^2 + 4 \times 10^1 + 3 \times 10^0 + 5 \times 10^{-1} \tag{7.2}$$

which will be written as $(1943.5)_{10}$ when there is a question about the base of the number system. (The exponent and base will always be expressed in the decimal number system.) The three number systems of primary importance to digital electronics are the decimal (base = 10), binary (base = 2), and hexadecimal (base = 16). Hexadecimal is simply a short hand way of representing binary numbers; had the human race evolved with 16 fingers and toes the topic of digital electronics would be vastly simpler. Of course, things could always be worse which would have occurred if the human body wasn't symmetric and common arithmetic was based on an odd numbered base.

Binary numbers can be converted to decimals numbers by using Eq. 1. For instance,

$$\begin{aligned}(100110101)_2 &= 1 \times 2^8 + 0 \times 2^7 + 0 \times 2^6 + 1 \times 2^5 \\ &+ 1 \times 2^4 + 0 \times 2^3 + 1 \times 2^2 + 0 \times 2^1 + 1 \times 2^0 = (309)_{10}\end{aligned} \tag{7.3}$$

Since there are only two symbols in the binary number system it requires a larger number of binary digits to represent a number than in the decimal system. For instance, the decimal number $(309)_{10}$ requires only three decimal digits but nine binary digits. A binary digit is known as a bit. An eight bit binary word is known as a byte and a four bit binary word is known as a nibble.

A decimal number may be converted to binary by dividing the decimal number by 2 and recording the quotient and remainder. This quotient is then divided by 2 and its remainder recorded. The process is continued until the quotient is 0. The remainders are the binary representation of the number. For instance, to convert the decimal number $(309)_{10}$ to binary

	Quotient	Remainder	
309/2 =	154	1	LSB (Least Significant Bit)
154/2 =	77	0	
77/2 =	38	1	
38/2 =	19	0	
19/2 =	9	1	
9/2 =	4	1	
4/2 =	2	0	
2/2 =	1	0	
1/2 =	0	1	MSB (Most Significant Bit)

which verifies Eq. 7.3.

The hexadecimal number system uses base 16. It uses the same symbols as the decimal number system for the first 10 symbols, i.e. 0,1,2,3,4,5,6,7,8,9. The next six symbols are the first six letters of the English alphabet, i.e. A, B, C, D, E, and F which represent the decimal numbers 10, 11, 12, 13, 14, and 15 respectively. The great utility of the hexadecimal number system is its simple relationship to the binary number system. The hexadecimal representation of a number can be obtained by grouping the binary digits in groups of four from the LSB to the MSB and replacing each four bit group with its hexadecimal equivalent. For instance the decimal number 309 is given by

$$(309)_{10} = (100110101)_2 = 0001\ 0011\ 0101 = (135)_H \tag{7.4}$$

where the symbol "H" is often used for hexadecimal. Thus, the hexadecimal number system can be thought of as simply a short hand method of representing binary numbers. Only 3 hexadecimal digits were required to represent the decimal number 309 whereas it required 9 binary digits. Only 3 digits would be required to represent this number is the decimal number system but the conversion from decimal to binary is not nearly as simple as the conversion from hexadecimal to binary.

Memory sizes in computers are usually specified in terms of bytes where a byte represents 8 bits. Since memories tend to be relatively large, the units that are used are 1 K byte $= 2^{10} = 1,024$ bytes, 1 M byte $= 2^{20} = 1,048,576$ bytes, and 1 G byte $= 2^{30} = 1,073,741,824$ bytes. For instance 16 M bytes of memory could be specified with memory addresses ranging from $(00000)_H$ to $(FFFFF)_H$ which is much more compact than the equivalent binary addresses which would be from 20 0's to 20 1's. The memory addresses corresponding to 1/4, 1/2, and 3/4 through this memory would be given by $(3FFFF)_H$, $(7FFFF)_H$, and $(BFFFF)_H$ respectively.

Computers must not only be capable of storing and processing positive integers but negative integers and floating point numbers. To differentiate between positive and negative integers signs such as the familiar "+" and "−" are used in ordinary arithmetic. With a computer only the symbols "1" and "0" are available.

One common way of representing negative integers is to use the most significant digit as a sign digit. A zero is used as the sign digit for positive integers and a one is used for negative integers. This system which is known as the sign-magnitude has the disadvantage of having both a positive and negative zero and is unwieldy for mathematical computations.

Another method of representing negative integers is to represent them as the 2's complement of the corresponding positive integer. The 2's complement of a binary number is obtained by first forming the 1's complement of this number and adding a "1" to the 1's complement. The 1's complement is obtained by exchanging all the 0's and 1's of the number to be complemented (including the sign bit). This system has the advantage of having only one representation of zero and mathematical operations are greatly simplified because the same circuits can be used for binary addition and subtraction. The 2's complement of a number is simply the number that would have to be added to it to obtain a sum of zero with a carry of one.

A third method of representing negative integers is known as offset binary. This system assigns the binary number consisting of all 0's to the most negative integer to be represented and all 1's to the most positive integer to be represented. Numbers in between are represented by an increasing sequence of binary numbers. This system is commonly employed in analog-to-digital converters since it permits a binary counter or register to be used in the conversion process.

The different binary representations for the decimal integers from −8 to +7 are shown in Table I.

Decimal Integer	Sign-Magnitude	Offset Binary	2's Complement
+7	0111	1111	0111
+6	0110	1110	0110
+5	0101	1101	0101
+4	0100	1100	0100
+3	0011	1011	0011
+2	0010	1010	0010
+1	0001	1001	0001
+0	0000	1000	0000
-0	1000	......	......
-1	1001	0111	1111
-2	1010	0110	1110
-3	1011	0101	1101
-4	1100	0100	1100
-5	1101	0011	1011
-6	1110	0010	1010
-7	1111	0001	1001
-8		0000	1000

Table I. 4 Bit Representation of Signed Integers

Note that the 4 bit 2's complement and offset binary can represent integers from −8 to +7 while the sign-magnitude can only represent integers from −7 to +7 because it has two representations of zero. Also note that the 2's complement can be obtained from the offset binary by simply complementing the most significant bit which makes the conversion from one to the other particularly simple.

Floating point numbers are represented in computers using a form of the sign magnitude. This consists of a sign bit followed by the exponent and then the mantissa. This complicates matters somewhat but computers can quickly and efficiently process such numbers.

7.2.4 Decimal Codes

Although computers and other digital systems prefer binary numbers, homo sapiens don't. Numbers are usually entered into computers in digital form. Additionally, other symbols such as letters of the English and Greek alphabet, spaces, carriage returns, etc. must be entered into and retrieved from digital systems. Since the only symbols used in digital electronics are 1's and 0's which correspond to one of two voltage levels in a digital electronic circuit, a binary digital code must be used for the ensemble of symbols that are required.

Binary coded decimal is a way of using a binary code to represent decimal digits. This is accomplished by encoding each of the decimal digits in the number using a four bit binary code. The four bit binary code that is used corresponds to the binary representation of the integers from 0 to 9; the binary codes corresponding to the decimal integers 10 to 15 are not used. Each decimal digit in the number is replaced by its four bit binary code. This is best illustrated by an example. The decimal number $(1943)_{10}$ would be represented as

$$(1943)_{10} = (0001\ 1001\ 0100\ 0011)_{BCD} \tag{7.5}$$

in binary coded decimal (BCD) where a space has been inserted between each grouping of four binary digits for clarity. Note that this is not the binary number for $(1943)_{10}$. Indeed, the binary representation for this number is given by

$$(1943)_{10} = (797)_H = (11110010111)_2 \tag{7.6}$$

which requires only 11 binary digits whereas the binary coded decimal representation requires 12 binary digits. Thus BCD is a binary code for decimal numbers. The advantage of the BCD representation of a number is that it can easily be connected to a digital display device which will display the number in decimal or base 10 form.

BCD is normally used in counters. Counters are devices which count voltage pulses; they are also used in instruments known as frequency counters which count the number of times a voltage waveform crosses a preselected voltage level with either a positive or negative slope in one second. An N decimal digit counter would use N BCD counters. These N counters would be cascaded. The least significant bit BCD counter would count from 0 to 9 and the next input pulse would reset the counter to 0 and produce a count out (or ripple count out in the parlance of digital electronics). Each BCD counter produces an output count for each 10 input counts or pulses and resets itself to 0. The output of each BCD counter is then connected to a digital display device which decodes the binary code for the decimal digit and displays the decimal digit. Thus, the count is directly displayed in decimal form. This topic will be investigated in Experiment 7 where a digital display device known as a 7 segment display will be used.

For digital systems that require more than just numbers a more extensive code must be used. One of the most common is the ASCII (American Standard Code for Information Interchange). This uses an eight bit code to encode 128 symbols which include decimal digits, the upper and lower case letters in the English alphabet, standard punctuation, printer control symbols, etc. Seven of the eight bits are required for the code and the eighth bit is used for parity check.

Another alphanumeric code that is used is the EBCDIC (Extended BCD Interchange Code). It is simply an extension of the BCD code and uses 9 bits to encode the same symbols as ASCII. Eight code bits are required and the ninth bit is used as a parity check.

Another binary code that is used for decimal integers is the Gray code. This code requires the same number of bits as the binary number for the decimal digits to be represented. The salient feature of the Gray code is that an increment in the decimal integer produces a change in only one bit position for the Gray code. The Gray code is obtained by beginning with a binary code of all 0's for the decimal integer 0. The Gray code for the next decimal integer is obtained by changing the least significant bit that produces a change in only one bit position for the code and, of course, produces a new code word. A four bit Gray code is given in Table II along with the binary number for the decimal integer which is also known as the natural binary code or simply the binary code.

Decimal Integer	Binary Number	Gray Code
0	0000	0000
1	0001	0001
2	0010	0011
3	0011	0010
4	0100	0110
5	0101	0111
6	0110	0101
7	0111	0100
8	1000	1100
9	1001	1101
10	1010	1111
11	1011	1110
12	1100	1010
13	1101	1011
14	1110	1001
15	1111	1000

Table II. 4 Bit Binary and Gray Code

Gray codes are used in mechanical shaft-angle encoders as shown in Fig. 7.1. A fixed disk is attached to the shaft whose angular position is to be encoded. Lamps are used to illuminate the disk and photodetectors are placed on the other side of the disk. If no light is detected this is encoded as a binary "1" and if light is detected this is encoded as a binary "0". The outer ring corresponds to the LSB and the inner most of rings is the MSB. The 4 bit code that is shown can be used to encode 16 angular positions. (A larger code could obviously encode more angular positions.)

Fig. 7.1 shows a natural binary code disk and a Gray code disk. If the natural binary code were used and the disk were positioned exactly halfway between decimal positions 0 and 15 the photodetectors would receive half as much light as is needed to make a decision as to whether the code bit is a 0 or 1 and the probability that it would decode each bit correctly would be 50%. Thus with the binary code any angular position from 0 to 15 could be decoded with the shaft in this position which would result in utter chaos. With the Gray code the maximum error in the angular position of the shaft would be $360°/(16 \times 2) = 5.63°$ for any position.

7.2.5 Boolean Algebra, Truth Tables and Logic Gates

The mathematics of digital electronics is Boolean algebra. Variables are represented using letters of the alphabet such as A, B, C, F, etc. just as in ordinary algebra. However, Boolean variables can have only one of two values. Various names are given to these values such as true and false; 1 and 0; or high and low (this corresponds to the voltage levels that will be used in electronic circuits to represent these two variables).

There are three basic operations in Boolean algebra (which will be indicated by all capital letters): *NOT*, *AND*, and *OR*:

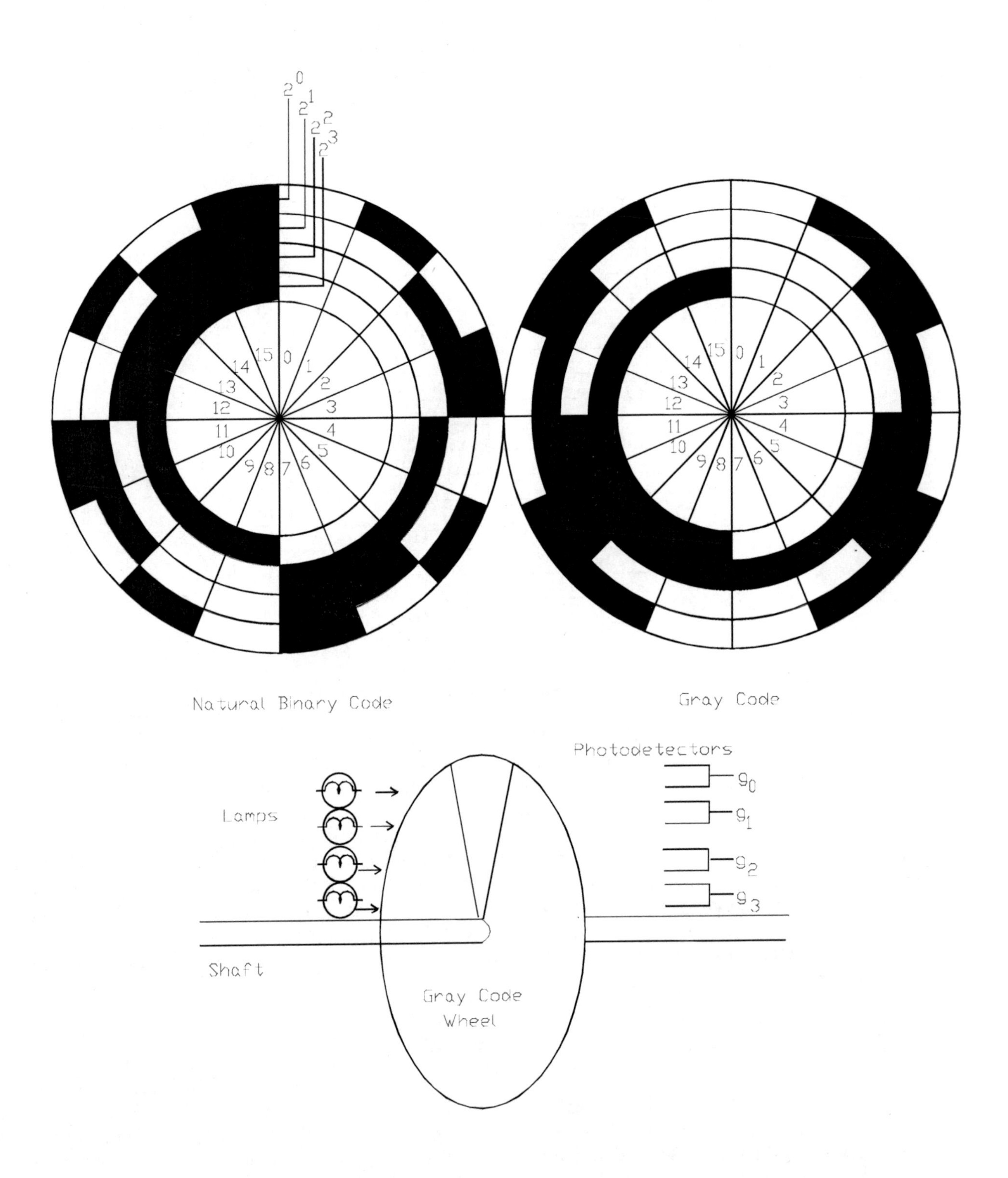

Figure 7.1: Gray wheel code.

NOT

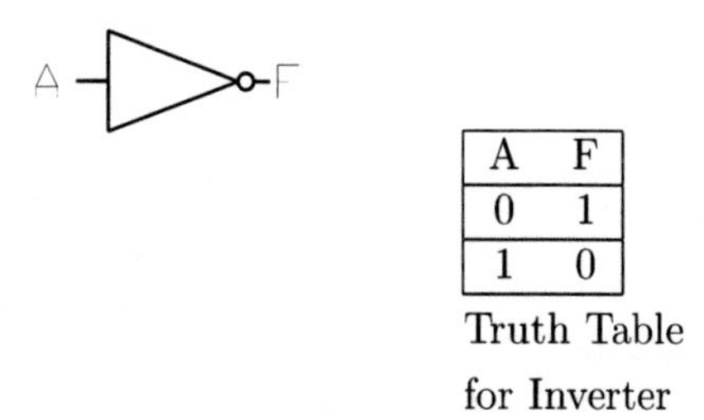

A	F
0	1
1	0

Truth Table
for Inverter

This is indicated by placing a bar or overscore over the variable, e.g. $F = \overline{A}$. This means that F is what A is not. If A is 0 then F is 1 and vice versa. This operation is also known as the complement. The circuit symbol for the logic gate or electronic circuit that performs this operation is shown above along with the truth table for this gate. The small circle on the end of the triangle indicates negation or complementation. This logic gate is also known as an $INVERTER$. A truth table is a combination of all the possible inputs and outputs for a digital logic gate or circuit. Since there is one input the truth table has $2^1 = 2$ entries.

AND

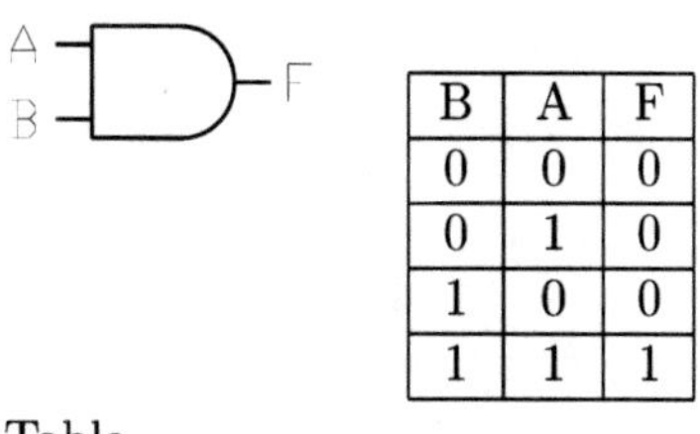

B	A	F
0	0	0
0	1	0
1	0	0
1	1	1

Truth Table
for 2-Input
AND Gate

The circuit symbol for the 2 input AND logic gate and its truth table are given above. This operation is represented as $F = AB$ for a two input AND gate; AND gates may have any number of inputs. This means that for a two input AND gate F is 1 if and only if both A and B are 1; otherwise it is 0. Although this appears to be simply the product of the two binary numbers A and B it should be interpreted as the logical AND operation and nothing else. Note that since there are two inputs that this truth table has $2^2 = 4$ lines. Also note that A and B are listed as though they represented the two digits of a two bit binary number.

Shown below are the circuit symbol and truth table for a 3 input AND gate for which $F = ABC$. The output is one if and only if all the inputs are one. If any input is zero, the output is zero. Since there are three inputs, the truth table has $2^3 = 8$ lines.

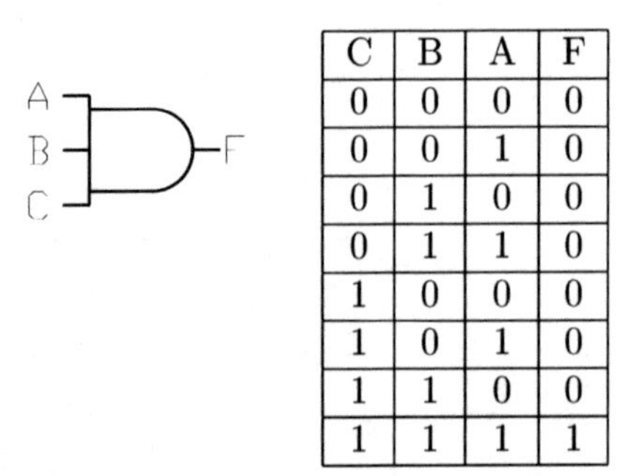

C	B	A	F
0	0	0	0
0	0	1	0
0	1	0	0
0	1	1	0
1	0	0	0
1	0	1	0
1	1	0	0
1	1	1	1

Truth Table for
3-Input *AND* Gate

OR

The two input *OR* gate is represented by $F = A + B$; *OR* gates may have any number of inputs. The plus operator " + " implies a logical *OR* operation and not the addition of two numbers. The *OR* operation produces an output of 1 if either A or B or both are 1; it is 0 if and only if both A and B are 0. This implies that $1 + 1 = 1$ so this is certainly not addition. The truth table and circuit symbol for a two input *OR* gate are shown below.

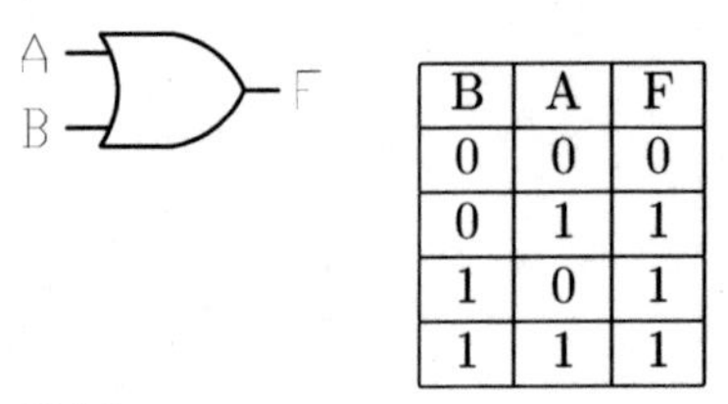

B	A	F
0	0	0
0	1	1
1	0	1
1	1	1

Truth Table
for 2-Input
OR Gate

A three input *OR* gate would be specified by $F = A + B + C$. The variable F would be 1 if any or all of the three input variables or any two of the variables A, B, or C are 1. The output is zero if and only if all three inputs are zero. The truth table and circuit symbol for the three input *OR* gate are shown below.

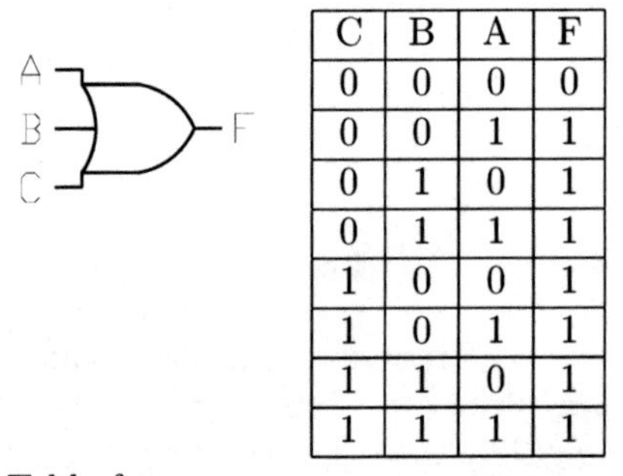

C	B	A	F
0	0	0	0
0	0	1	1
0	1	0	1
0	1	1	1
1	0	0	1
1	0	1	1
1	1	0	1
1	1	1	1

Truth Table for
3-Input *OR* Gate

In addition to the three basic operations of *NOT*, *AND*, and *OR*, there are so other operations that are combinations of the basic three that occur so often that they are given special names. These are the $NAND$ gate, the NOR gate, and the $EXCLUSIVE\ OR$ gate. The $NAND$ gate is obtained by cascading an $INVERTER$ with the output of an AND gate. Similarly, a NOR gate is obtained by cascading an $INVERTER$ with an OR gate. Finally, an $EXCLUSIVE\ OR$ gate is obtained by the function $F = A\overline{B} + \overline{A}B = A \oplus B$; the symbol "$\oplus$" will be used to indicate the $EXCLUSIVE\ OR$ function.

$NAND$

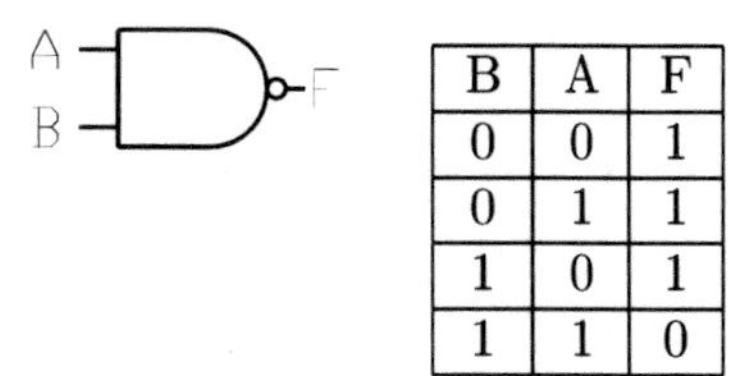

B	A	F
0	0	1
0	1	1
1	0	1
1	1	0

Truth Table
for 2-Input
$NAND$ Gate

The circuit symbol and truth table for a two input $NAND$ gate are shown above. The small circle on the output of the symbol indicates negation or complementation. This operation is given by $F = \overline{AB}$. (Note that the bar extends over both symbols.) The output is 0 if and only if both A and B are 1; otherwise the output is 1.

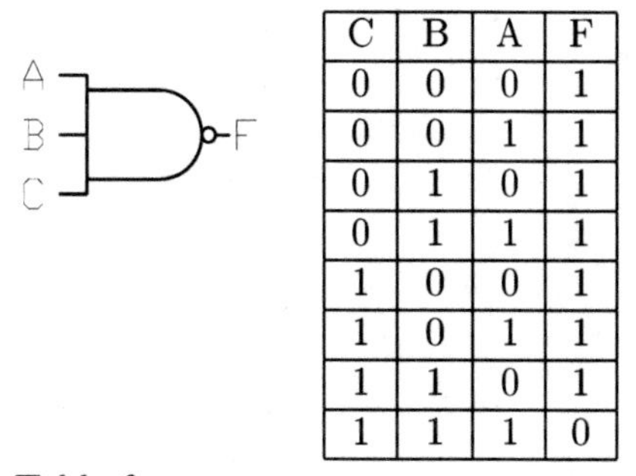

C	B	A	F
0	0	0	1
0	0	1	1
0	1	0	1
0	1	1	1
1	0	0	1
1	0	1	1
1	1	0	1
1	1	1	0

Truth Table for
3-Input $NAND$

Shown above are the circuit symbol and truth table for the three input $NAND$ gate. This operation is given by $F = \overline{ABC}$. (Note that the bar extends over all 3 symbols.) Note that the output is 0 if and only if all the inputs are 1; otherwise the output is 1.

NOR

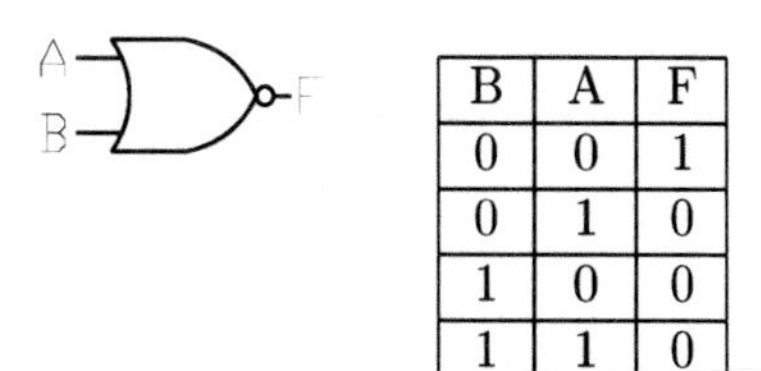

B	A	F
0	0	1
0	1	0
1	0	0
1	1	0

Truth Table
for 2_Input
NOR

A *NOR* gate is obtained by cascading an *OR* gate with an *INVERTER*. The circuit symbol and truth table for a 2 input *NOR* gate are shown above. This operation is given by $F = \overline{A + B}$. Note that the output is 1 if and only if both inputs are 0; otherwise the output is 0.

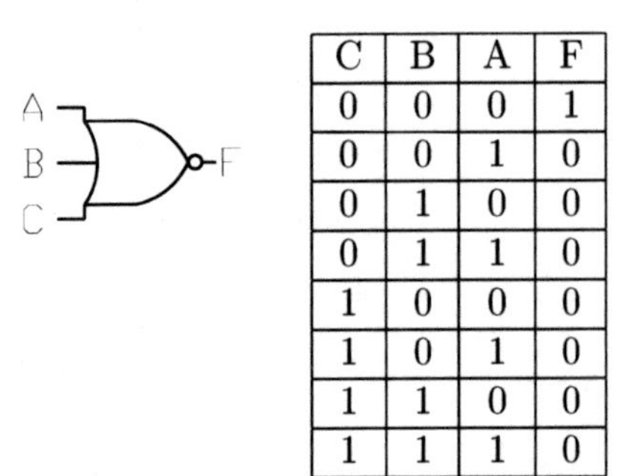

C	B	A	F
0	0	0	1
0	0	1	0
0	1	0	0
0	1	1	0
1	0	0	0
1	0	1	0
1	1	0	0
1	1	1	0

Truth Table for
3-Input *NOR*

The circuit symbol and truth table for a three input *NOR* gate are shown above. This operation is given by $F = \overline{A + B + C}$. Note that the output is 1 if and only if all three inputs are 0; otherwise the output is 0.

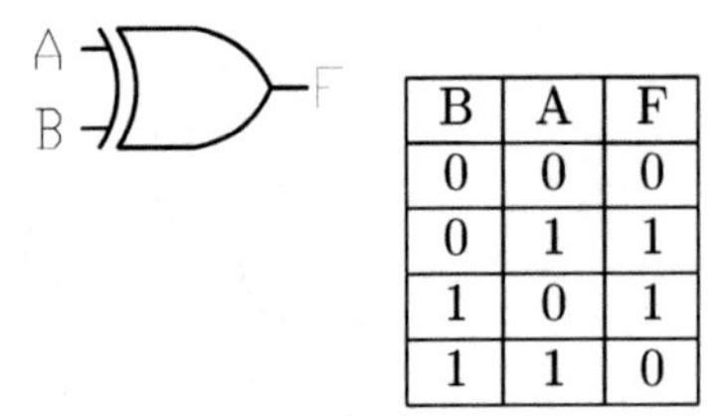

B	A	F
0	0	0
0	1	1
1	0	1
1	1	0

Truth TAble
for 2-Input
EXOR

The circuit symbol and truth table for a two input *EXCLUSIVE OR* gate are shown above. This operation is given by $F = A \oplus B = \overline{A}B + A\overline{B}$. Note that the output is 0 if both inputs are the same and 1 is the inputs are different.

Each of these logic gates are known as combinational logic gates since the output is only a function of the input. In subsequent experiments other types of logic gates will be examined for which the output is a function of both the inputs and the current state of the gate; these are known as sequential logic gates.

7.2.6 Boolean Identities

The most important identities of Boolean algebra are:

$$A + 0 = A \quad A \cdot 1 = A \tag{7.7}$$

$$A + 1 = 1 \quad A \cdot 0 = 0 \tag{7.8}$$

$$A + A = A \quad A \cdot A = A \tag{7.9}$$

$$A + \overline{A} = 1 \quad A \cdot \overline{A} = 0 \tag{7.10}$$

$$A + B = B + A \quad AB = BA \tag{7.11}$$

$$A + (B + C) = (A + B) + C \quad A(BC) = (AB)C \tag{7.12}$$

$$A(B + C) = AB + AC \quad A + BC = (A + B)(A + C) \tag{7.13}$$

$$\overline{A + B} = \overline{A} \cdot \overline{B} \quad \overline{A \cdot B} = \overline{A} + \overline{B} \tag{7.14}$$

along with the identity operation $\overline{\overline{A}} = A$. Eq. 7.14 is particularly important and is known as DeMorgan's Theorem; it can be extended to any number of variables

$$\overline{A_1 + A_2 + \cdots + A_N} = \overline{A_1} \cdot \overline{A_2} \cdots \overline{A_N}$$

$$\overline{A_1 \cdot A_2 \cdot \cdots \cdot A_N} = \overline{A_1} + \overline{A_2} + \cdots + \overline{A_N} \tag{7.15}$$

which indicates that the complement of the *OR* of a number of variables is equal to the *AND* of their complements and vice versa.

7.2.7 Binary to Gray Code and Gray to Binary Code Converters

One application of *EXCLUSIVE OR* gates is to perform the conversion from binary to Gray or vice versa. These codes are given in Table II. The circuits necessary for the conversion are shown in Fig. 7.2.

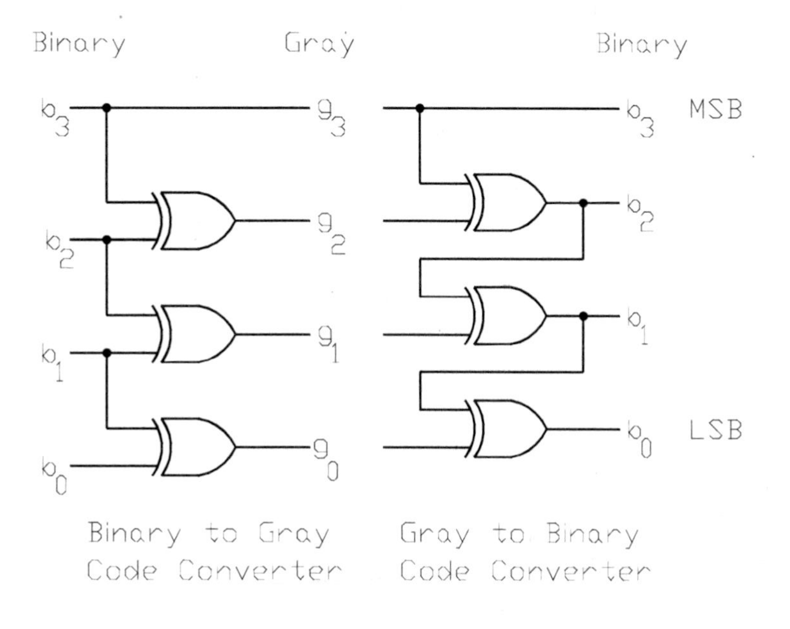

Figure 7.2: Binary to Gray and Gray to binary code converters.

7.2.8 Binary Addition

Binary logic gates can be used to add binary numbers. To add two binary integers, each of which has N bits, require 1 circuit known as a half adder and $N-1$ circuits known as full adders. These circuits are shown in Fig. 7.3. The half adder is used to add the LSB of the two binary numbers $(A_N...A_1)_2$ and $(B_N...B_1)_2$, i.e. A_1 and B_1.

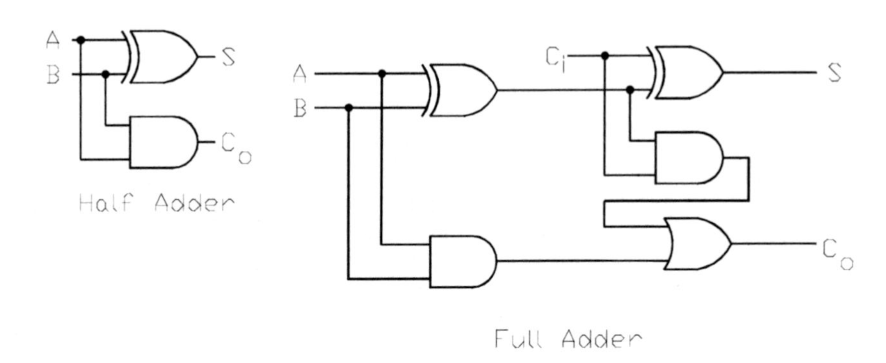

Figure 7.3: Half and full adders.

The half adders has two inputs: A_1 and B_1 and two outputs: S_1 (the sum output) and C_{01} (the carry output). (The 1 subscript has been omitted from the figure to simplify the figure.) The logic equations for the half adders can be obtained from Fig. 7.3

$$S_1 = A_1 \oplus B_1 \tag{7.16}$$

$$C_{o1} = A_1 B_1 \tag{7.17}$$

for which the truth table is given by

B_1	A_1	S_1	C_{01}
0	0	0	0
0	1	1	0
1	0	1	0
1	1	0	1

Truth Table for Half-Adder

The carry output for the half adder would be the carry in input for the full adder that would be used to add A_2 and B_2. Each stage produces a sum out, S_n, and a carry out, C_n. Each stage (after stage 1, the LSB) has three inputs A_n, B_n, and $C_{in} = C_{o,n-1}$.

The outputs of the full adder can be obtained from Fig. 7.3 as

$$S_n = C_{in} \oplus (A_n \oplus B_n) \tag{7.18}$$

$$C_{on} = A_n B_n + C_{in}(A_n \oplus B_n) \tag{7.19}$$

which are the sum and the carry out for the nth stage. The truth table for the full adder can be obtained from Fig. 7.3

A_n	B_n	C_{in}	C_{on}	S_n
0	0	0	0	0
0	0	1	0	1
0	1	0	0	1
0	1	1	1	0
1	0	0	0	1
1	0	1	1	0
1	1	0	1	0
1	1	1	1	1

Truth Table for Full Adder

Subtraction of two binary numbers can be accomplished using the same circuit as the adder by changing the number to subtracted using its 2's complement form. Other circuit exist for other arithmetic operations on binary numbers.

7.2.9 Logic Families

Digital electronics to implement logic function are available in an almost bewildering variety of families. A family indicates the type of electronic device used to implement the logic function and the density of the electronic devices on the integrated circuit. Computer microprocessors use VLSI integration for which there are thousands of transistors on a chip and the active devices are either NMOS or CMOS transistors.

The ICs that will be used for experiments in this course use medium scale integration which means that no more than 100 gates are contained in the packages. The most popular MSI ICs are the TTL and CMOS families. This course will use the 74LS TTL family where LS stands for low-power Schottky, TTL stands for transistor-transistor logic (which means that the logic functions are implement with BJTs), and the 74 prefix is a numbering systems for devices that was developed at Texas Instruments in the 1970s.

The specific devices to be used in this experiment are:
74LS00 Quad 2-Input $NAND$ Gate
74LS02 Quad 2-Input NOR Gate
74LS04 Hex $INVERTER$ Gate
74LS08 Quad 2-Input AND Gate
74LS11 Triple 3-Input AND Gate
74LS27 Triple 3-Input NOR Gate
74LS32 Quad 2-Input OR Gate
74LS86 Quad 2-Input $EXCLUSIVE\ OR$ Gate where the first column is the part number and the second column indicates the logic function, the number of inputs each gate has, and the number of gates on the IC.

Pin diagrams for each of these devices are contained at the end of the laboratory manual.

Pin 14 of each IC is to be connected to a $+5\ V$ DC supply and pin 7 is to be connected to ground. Each of these devices in the TTL 74LS family operates with a power supply voltage of $+5\ V$; it will tolerate power supply voltages up to 7 volts but if a larger voltage is used the IC will be irreversibly damaged, i.e. incinerated.

Each member of the 74LS family will interpret an input voltage from $-0.2\ V$ to $0.4\ V$ as a low (logical 0) and an input voltage from $2\ V$ to $5.25\ V$ as a high (logical 1). Each member of this family will produce a low output that ranges from 0 to $0.2\ V$ and a high output that ranges from 2.5 to 5 Volts. Voltages that do not lie in these ranges will not be properly interpreted.

7.3 Procedure

7.3.1 Familiarization

Observe the digital ICs that will be used in this experiment. They appear to be small gray or black bugs with 14 legs. The numbering scheme used for these ICs is the same as that used for the op-amps that were used in previous experiments. Namely, there is a "U shaped" indentation on one end and pin 1 is to left of this; there may also be a small circular indentation by pin 1. (A large circle in the center of the case doesn't have anything to do with the pin numbering system.) Pin 14 is directly opposite pin 1. These ICs are TTL 74LS combinational logic gates. The specific gates that will be used in this experiment are:

74LS00 Quad 2-Input $NAND$ Gate

74LS02 Quad 2-Input NOR Gate

74LS04 Hex $INVERTER$ Gate

74LS08 Quad 2-Input AND Gate

74LS11 Triple 3-Input AND Gate

74LS27 Triple 3-Input NOR Gate

74LS32 Quad 2-Input OR Gate

74LS86 Quad 2-Input $EXCLUSIVE\ OR$ Gate where the first column is the part number and the second column indicates the logic function, the number of inputs each gate has, and the number of gates on the IC. Pin diagrams for each of these devices are contained at the end of the laboratory manual. Since each IC is preceded by a prefix of 74LS, this prefix will often be omitted in the diagrams and replaced with an apostrophe.

Pin 14 of each IC is to be connected to a +5 V DC supply and pin 7 is to be connected to ground. Each of these devices in the TTL 74LS family operates with a power supply voltage of +5 V; it will tolerate power supply voltages up to 7 volts but if a larger voltage is used the IC will be irreversibly damaged, i.e. incinerated. If the power supply connections are reversed, it will similarly be irreversibly destroyed.

Note the logic switches at the lower left of the CADET. There are 8 switches which will be used as inputs to the ICs. There is a switch to the left of the bus for these ICs which should be set to the +5 V position rather than $+V$. The logic switches will be referred to in the following experiment as LSN where LS stands for logic switch and N is an integer from 1 to 8. The 1 position for these two position switches is +5 V (logic 1) and the 0 position is 0 V (logic 0).

Note the LED logic state indicators on the right of the CADET. The switch below the logic state indicators should be set to TTL rather than CMOS and the voltage level switch to the left of the indicators should be set to +5 V rather than $+V$. The red LED represents a high (+5 V or binary 1) and the green LED represents a low (0 V or binary 0). The logic state indicators will be referred to as LIN where LI stands for logic indicator and N is an integer from 1 to 8.

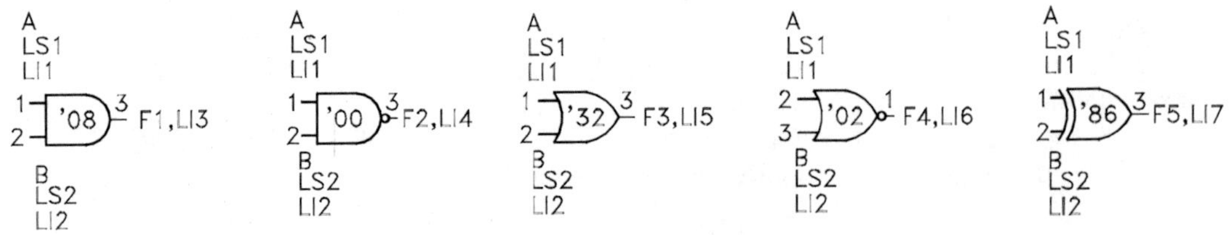

Figure 7.4: Two-input logic gates.

7.3.2 Truth Table for Two Input Logic Gates

Assemble the circuit shown in Fig. 7.4 with the power off. Begin by establishing busses for +5 V and the system ground. Connect pin 14 of each of the 5 ICs to the +5 V supply. Connect pin 7 of each of the 5 ICs to the system ground. Each of these two input ICs will have LS1 and LS2 as their A and B inputs and logic indicators LI1 and LI2 will be used to monitor the A and B inputs. Note that for some perverse reason that the pin numbering used on the 74LS02 NOR gate differs from that of the other 4 ICs. Turn the power on. Complete the following truth table for the two input combinational logic gates.

A LS1 LI1	B LS2 LI2	'08 AND LI3	'00 NAND LI4	'32 OR LI5	'02 NOR LI6	'86 EXOR LI7
0	0					
0	1					
1	0					
1	1					

Truth Table for Two-Input Logic Gates

Demonstrate the properly functioning circuit to the laboratory instructor.

Laboratory Instructor Verification ____________________

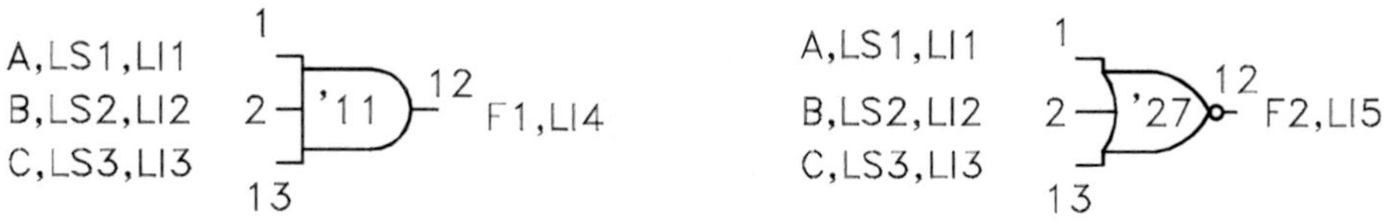

Figure 7.5: Three-input logic gates.

7.3.3 Truth Table for 3-Input Logic Gates

Assemble the circuit shown in Fig. 7.5 with the power off. Connect pin 14 to the +5 V supply and pin 7 to ground for each IC. Turn the power on. Complete the truth table for the three input logic gates.

A LS1 LI1	B LS2 LI2	C LS3 LI3	'11 AND LI4	'27 NOR LI5
0	0	0		
0	0	1		
0	1	0		
0	1	1		
1	0	0		
1	0	1		
1	1	0		
1	1	1		

Truth Table for 3-Input Logic Gates

Demonstrate the properly functioning circuit to the laboratory instructor.

Laboratory Instructor Verification ______________________________

Figure 7.6: DeMorgan's theorem for two variables (complement of sum).

7.3.4 DeMorgan's Theorem for Two Variables (Complement of Sum)

Assemble the circuit shown in Fig. 7.6 with the power off. Connect pin 14 of each IC to the +5 V supply and pin 7 to the system ground. Turn the power on. Obtain the truth table for the variables F1 and F2.

A LS1 LI1	B LS2 LI2	F1 LI3	F2 LI4
0	0		
0	1		
1	0		
1	1		

DeMorgan's Theorem for Two Variables— Complement of Sum

Demonstrate the properly functioning circuit to the laboratory instructor.

Laboratory Instructor Verification ______________________________

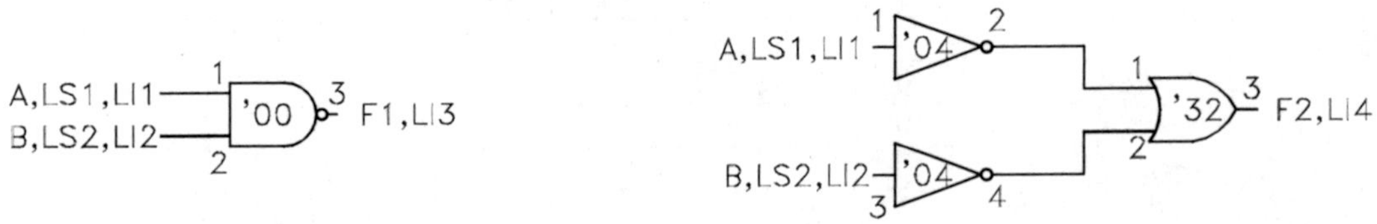

Figure 7.7: DeMorgan's theorem for two variables (complement of product).

7.3.5 DeMorgan's Theorem for Two Variables (Complement of Product)

Assemble the circuit shown in Fig. 7.7 with the power off. Connect pin 14 of each IC to the +5 V supply and pin 7 to ground. Turn the power on. Obtain the truth table for F1 and F2.

A LS1 LI1	B LS2 LI2	F1 LI3	F2 LI4
0	0		
0	1		
1	0		
1	1		

DeMorgan' Theorem for Two Variables—Complement of Product

Demonstrate the properly functioning circuit to the laboratory instructor.

Laboratory Instructor Verification ________________________________

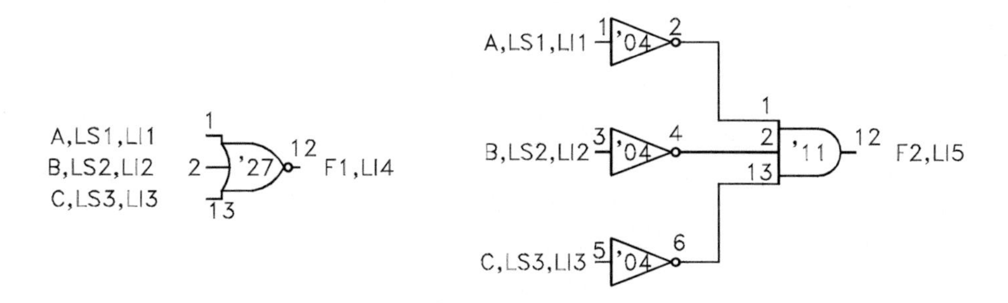

Figure 7.8: DeMorgan's theorem for three variables.

7.3.6 DeMorgan's Theorem for Three Variables

Assemble the circuit shown in Fig. 7.8 with the power off. Connect pin 14 of each IC to the +5 V supply and pin 7 to the system ground. Turn the power on. Obtain the truth table for F1 and F2.

A LS1 LI1	B LS2 LI2	C LS3 LI3	F1 LI4	F2 LI5
0	0	0		
0	0	1		
0	1	0		
0	1	1		
1	0	0		
1	0	1		
1	1	0		
1	1	1		

DeMorgan's Theorem for Three Variables

Demonstrate the properly functioning circuit to the laboratory instructor.

Laboratory Instructor Verification ____________________

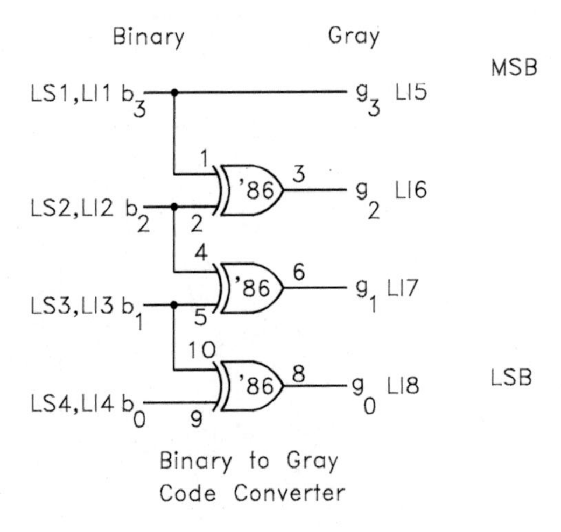

Figure 7.9: Binary to Gray code converter.

7.3.7 Binary to Gray Code Converter

Assemble the circuit shown in Fig. 7.9 with the power off. Connect pin 14 to the +5 V supply and pin 7 to the system ground. Turn the power on. Obtain the truth table for the Binary to Gray code converter.

b_3(LI1)	b_2(LI2)	b_1(LI3)	b_0(LI4)	g_3(LI5)	g_2(LI6)	g_1(LI7)	g_0(LI8)
0	0	0	0				
0	0	0	1				
0	0	1	0				
0	0	1	1				
0	1	0	0				
0	1	0	1				
0	1	1	0				
0	1	1	1				
1	0	0	0				
1	0	0	1				
1	0	1	0				
1	0	1	1				
1	1	0	0				
1	1	0	1				
1	1	1	0				
1	1	1	1				

Binary to Gray Code Converter

Demonstrate the properly functioning circuit to the laboratory instructor.

Laboratory Instructor Verification 

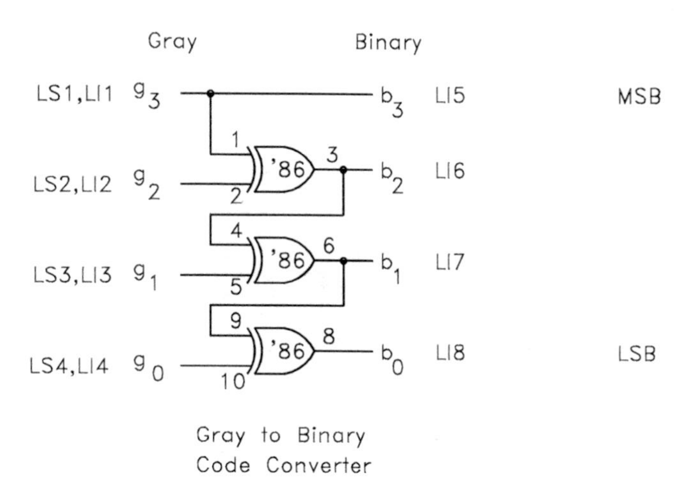

Figure 7.10: Gray to binary code converter.

7.3.8 Gray to Binary Code Converter

Assemble the circuit shown in Fig. 7.10 with the power off. Connect pin 14 to the +5 V supply and pin 7 to the system ground. Turn the power on. Obtain the truth table for the Gray to Binary code converter.

g_3(LI1)	g_2(LI2)	g_1(LI3)	g_0(LI4)	b_3(LI5)	b_2(LI6)	b_1(LI7)	b_0(LI8)
0	0	0	0				
0	0	0	1				
0	0	1	0				
0	0	1	1				
0	1	0	0				
0	1	0	1				
0	1	1	0				
0	1	1	1				
1	0	0	0				
1	0	0	1				
1	0	1	0				
1	0	1	1				
1	1	0	0				
1	1	0	1				
1	1	1	0				
1	1	1	1				

Gray to Binary Code Converter

Demonstrate the properly functioning circuit to the laboratory instructor.

Laboratory Instructor Verification ____________________

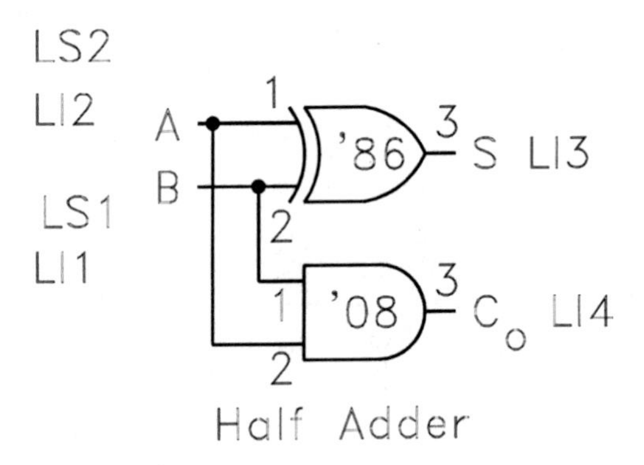

Figure 7.11: Half adder.

7.3.9 Half Adder

Assemble the circuit shown in Fig. 7.11 with the power off. Connect pin 14 of each IC to the +5 V supply and pin 7 to the system ground. Turn the power on. Obtain the truth table for the half adder.

B LS1 LI1	A LS2 LI2	S LI3	C_O LI4
0	0		
0	1		
1	0		
1	1		

Half Adder

Demonstrate the properly functioning circuit to the laboratory instructor.

Laboratory Instructor Verification ________________________________

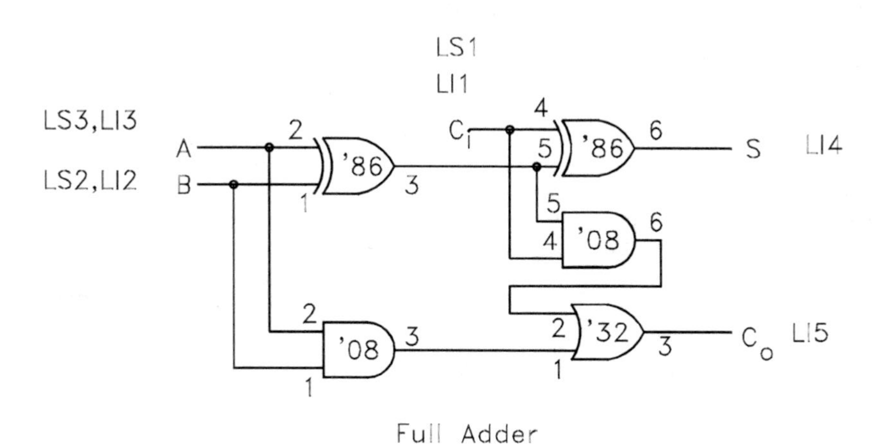

Figure 7.12: Full adder.

7.3.10 Full Adder

Assemble the circuit shown in Fig. 7.12 with the power off. Connect pin 14 of each IC to the +5 V supply and pin 7 to the system ground. Turn the power on. Obtain the truth table for the full adder.

C_i LS1 LI1	B LS2 LI2	A LS3 LI3	S LI4	C_o LI5
0	0	0		
0	0	1		
0	1	0		
0	1	1		
1	0	0		
1	0	1		
1	1	0		
1	1	1		

Full Adder

Demonstrate the properly functioning circuit to the laboratory instructor.

Laboratory Instructor Verification ______________________________

7.4 Laboratory Report

1. Turn in all truth tables that were recorded.
2. Compare the truth tables obtained for F1 and F2 in steps 7.3.4, 7.3.5, and 7.3.6. Explain the relationship between F1 and F2.

____

____

____

3. Draw a circuit that would implement a 7 bit Binary to Gray code conversion. How many rows would the truth table for this circuit require?

____

4. How many gates would be required to add two 7 bit binary numbers using the circuits examined in this experiment?

____

5. What are the birthdays of each member of the laboratory group in hexadecimal? (Do not drop the prefix '19' or '20'.)

____

6. Answer any supplementary questions that may have been posed by the laboratory instructor.

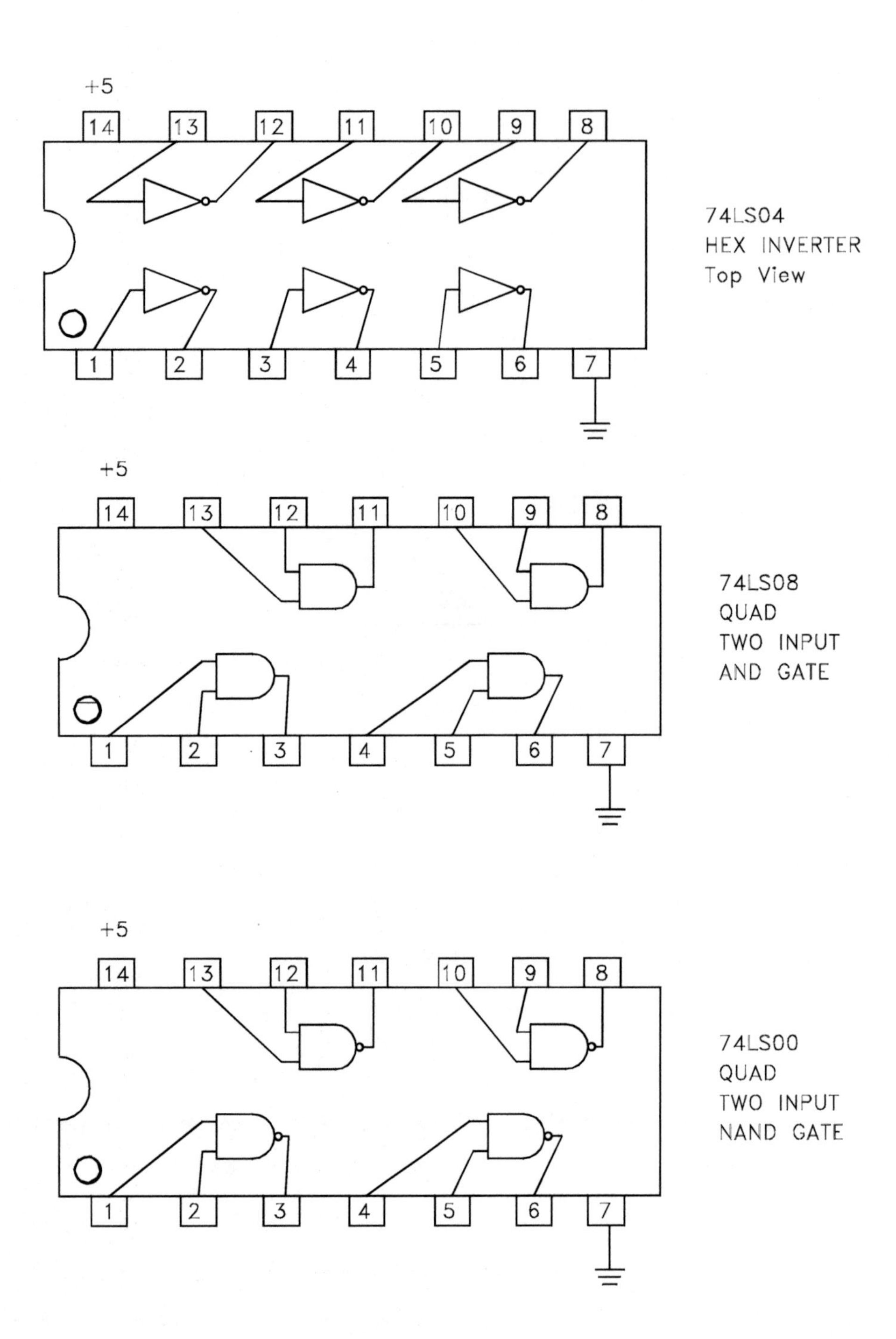
+5
14
13
12
11
10
9
8
1
2
3
4
5
6
7
74LS04
HEX INVERTER
Top View
+5
14
13
12
11
10
9
8
1
2
3
4
5
6
7
74LS08
QUAD
TWO INPUT
AND GATE
+5
14
13
12
11
10
9
8
1
2
3
4
5
6
7
74LS00
QUAD
TWO INPUT
NAND GATE

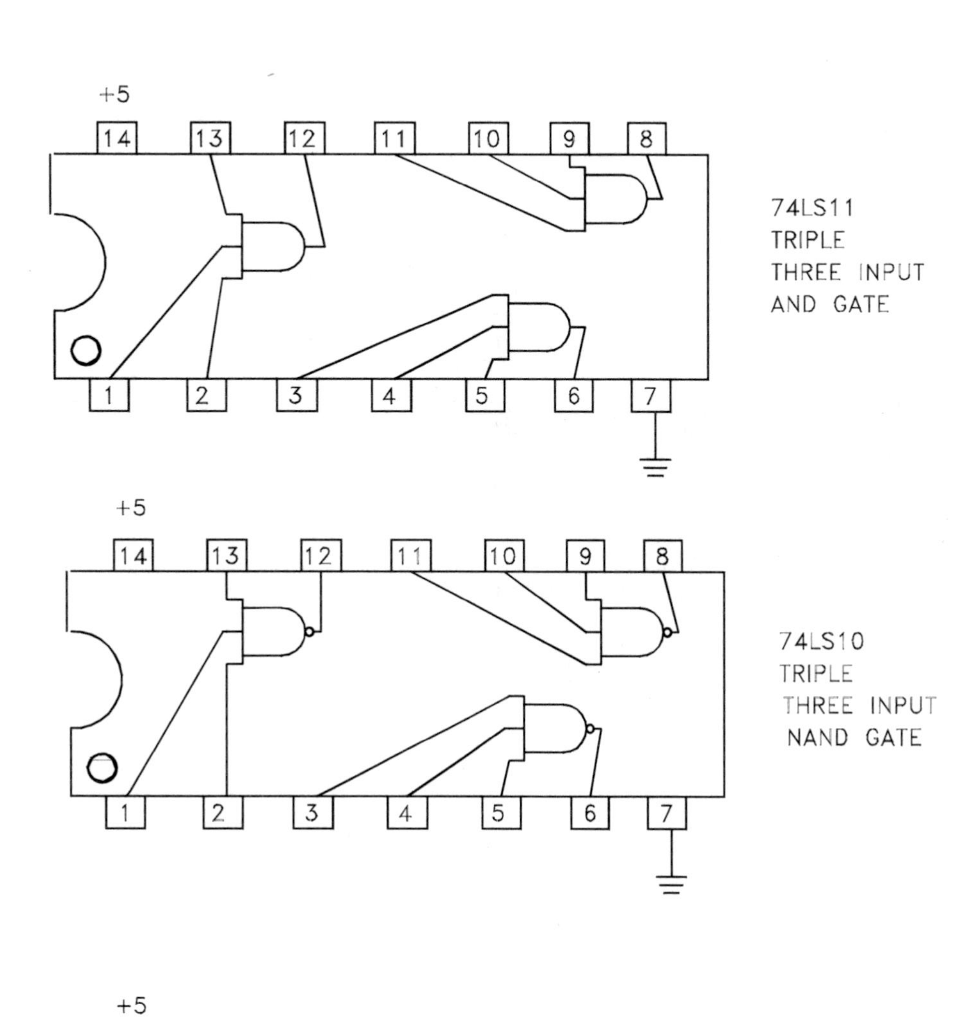

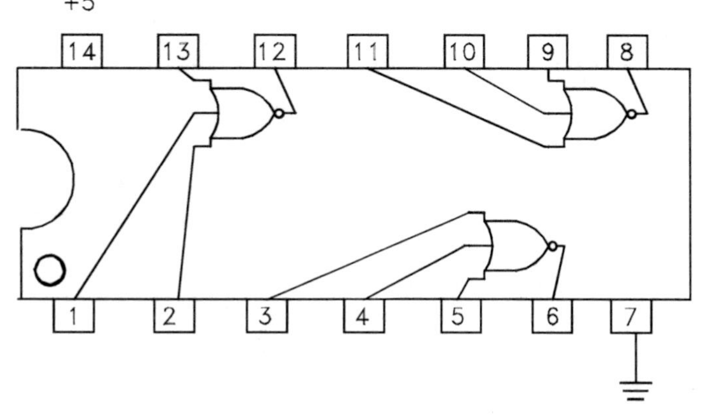

74LS27
TRIPLE
THREE INPUT
NOR GATE

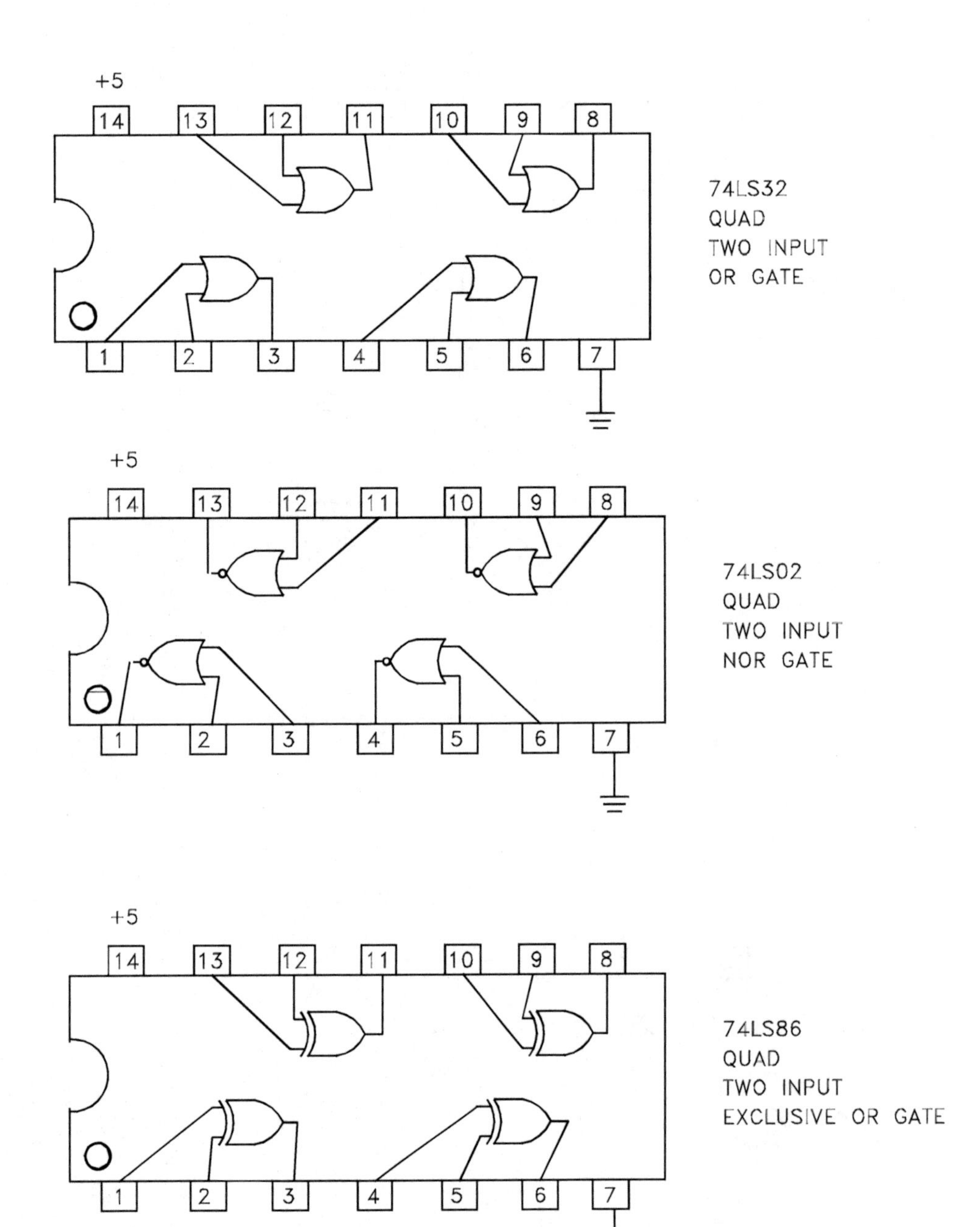
+5
14
13
12
11
10
9
8
1
2
3
4
5
6
7
74LS32
QUAD
TWO INPUT
OR GATE
+5
14
13
12
11
10
9
8
1
2
3
4
5
6
7
74LS02
QUAD
TWO INPUT
NOR GATE
+5
14
13
12
11
10
9
8
1
2
3
4
5
6
7
74LS86
QUAD
TWO INPUT
EXCLUSIVE OR GATE

Chapter 8

Logic Function Minimization and Implementation

8.1 Objective

The objectives of this experiment are to experimentally examine the minimization and implementation of logical functions of 3 and 4 Boolean variables. Karnaugh map minimization will be employed for function minimization. Function implementation will be achieved using the elementary 2 and 3 input combinational logic gates as well as multiplexers.

8.2 Theory

Digital design for combinational logic consists of clearly expressing the problem in words, determining the truth table or logic function from the problem statement, and then implementing the logical function with logic gates. This is a straight forward procedure for which unambiguously stating the problem is often the most difficult step.

Once the function has been obtained, it is best to see if it can be simplified before trying to directly implement it with hardware. For instance the function F of the four Boolean variables W, X, Y, and Z

$$F = F(W,X,Y,Z) = W \cdot X + \overline{W} \cdot X \cdot \overline{Y} + W \cdot \overline{X} \cdot Z + \overline{W} \cdot Y \tag{8.1}$$

would require 3 $INVERTER$s, 2 3-input AND gates, 2 2-input AND gates, and 1 4-input OR gates; this could be done without difficulty. However, using Boolean identities the function of Eq. 8.1 can be simplified as

$$F = F(W,X,Y,Z) = X + W \cdot Z + Y \cdot \overline{W} \tag{8.2}$$

which is considerably simpler than Eq. 8.1 since it requires only 2 2-input AND gate, 1 3-input OR gate, and 1 $INVERTER$. Thus it is best to examine logical functions to determine if a simplification is possible before implementing it with digital hardware.

As an example of digital design for combinational logic a control unit for a simple home security system will be considered. The system has three sensors: a door and window sensor, a smoke detector, and a motion sensor. The sensors for the windows and doors will use reed switches which are switches which are closed when close to a magnet. The magnets are place on the windows and doors and the reed switches are located on the door frames and window sills so that when they are closed the magnets are positioned next to the switches. The sensors for all the doors and windows to be protected are connected in series so that if any of them are opened the circuit is broken and the input to the control unit changes from a logical 0 to a logical

1. The motion detector is to change its output from a logical 0 to a logical 1 if motion is detected within its scan range. The smoke detector is to change its output from a logical 0 to a logical 1 if smoke is detected.

The control unit is to have a switch to arm the home security system. If the system is armed, and the motion detector and/or the door or window detectors are activated, an alarm is to sound and a telephone call placed to the monitoring service. If the smoke detector detects smoke, an alarm is to sound and the monitoring service called whether the control unit is armed or not.

The first step is to determine the Boolean variables. The obvious choices are:

$A \equiv$ the home security system is armed

$D \equiv$ the window and/or door detectors have detected an entry

$M \equiv$ the motion detector has detected motion

$S \equiv$ the smoke detector has detected smoke where the variables are 1 if the condition is true and 0 if false. The alarm is then a function of 4 Boolean variables. If F is the alarm function, then it is given by $F(A, D, M, S)$. The next step is to determine the truth table for this problem.

Row No.	A	D	M	S	F
0	0	0	0	0	0
1	0	0	0	1	1
2	0	0	1	0	0
3	0	0	1	1	1
4	0	1	0	0	0
5	0	1	0	1	1
6	0	1	1	0	0
7	0	1	1	1	1
8	1	0	0	0	0
9	1	0	0	1	1
10	1	0	1	0	1
11	1	0	1	1	1
12	1	1	0	0	1
13	1	1	0	1	1
14	1	1	1	0	1
15	1	1	1	1	1

Truth Table for Home Security System

The first column is the decimal number for the row; it begins with row 0 so that the decimal number is equivalent to the binary number that would be obtained if (ADMS) were interpreted as a 4 bit binary number. Since F is a function of 4 Boolean variables, it requires a truth table with $2^4 = 16$ rows.

Now that the truth table has been determined the next step is to determine a Boolean expression for F as a function of the four Boolean variables A, D, M, and S. This can be done by forming a logical OR of all of the rows in the truth table for which the output is 1.

$$F = F(A, D, M, S) = \overline{A} \cdot \overline{D} \cdot \overline{M} \cdot S + \overline{A} \cdot \overline{D} \cdot M \cdot S + \overline{A} \cdot D \cdot \overline{M} \cdot S + \overline{A} \cdot D \cdot M \cdot S + A \cdot \overline{D} \cdot \overline{M} \cdot S$$

$$+A \cdot \overline{D} \cdot M \cdot \overline{S} + A \cdot \overline{D} \cdot M \cdot S + A \cdot D \cdot \overline{M} \cdot \overline{S} + A \cdot D \cdot \overline{M} \cdot S + A \cdot D \cdot M \cdot \overline{S} + A \cdot D \cdot M \cdot S \qquad (8.3)$$

which would require 4 $INVERTER$s, 11 4-input AND gates, and 1 11-input OR gate. Therefore, it would be wise to consider simplifying Eq. 8.3 if possible.

Eq. 8.3 is in a form known as a standard or canonical sum of products. Each term in the sum contains all four literals (a literal is a Boolean variable in either its complemented or uncomplemented form). Each term in this sum is known as a minterm because it contains all the literals for the Boolean function F. A short hand method of writing Eq. 8.3 is

$$F = F(A, D, M, S) = \sum m(1, 3, 5, 7, 9, 10, 11, 12, 13, 14, 15) \tag{8.4}$$

where the summation sign indicates a sum of minterms, the small m indicates minterms, and the decimal numbers in the parentheses are the decimal row numbers for which F is 1.

An alternative method of implementing the function would be to determine $\overline{F}$ as a function of the four Boolean variables and then to take its complement. This would yield

$$\overline{F} = \overline{F}(A, D, M, S) = \overline{A} \cdot \overline{D} \cdot \overline{M} \cdot \overline{S} + \overline{A} \cdot \overline{D} \cdot M \cdot \overline{S} + \overline{A} \cdot D \cdot \overline{M} \cdot \overline{S} + \overline{A} \cdot D \cdot M \cdot \overline{S} + A \cdot \overline{D} \cdot \overline{M} \cdot \overline{S} \tag{8.5}$$

for the complement of F. This can then be complemented to obtain an expression for F

$$\overline{\overline{F}} = \overline{\overline{A} \cdot \overline{D} \cdot \overline{M} \cdot \overline{S} + \overline{A} \cdot \overline{D} \cdot M \cdot \overline{S} + \overline{A} \cdot D \cdot \overline{M} \cdot \overline{S} + \overline{A} \cdot D \cdot M \cdot \overline{S} + A \cdot \overline{D} \cdot \overline{M} \cdot \overline{S}} =$$

$$\overline{(\overline{A} \cdot \overline{D} \cdot \overline{M} \cdot \overline{S})} \cdot \overline{(\overline{A} \cdot \overline{D} \cdot M \cdot \overline{S})} \cdot \overline{(\overline{A} \cdot D \cdot \overline{M} \cdot \overline{S})} \cdot \overline{(\overline{A} \cdot D \cdot M \cdot \overline{S})} \cdot \overline{(A \cdot \overline{D} \cdot \overline{M} \cdot \overline{S})} = \tag{8.6}$$

$$(A + D + M + S) \cdot (A + D + \overline{M} + S) \cdot (A + \overline{D} + M + S) \cdot (A + \overline{D} + \overline{M} + S) \cdot (\overline{A} + D + M + S) \tag{8.7}$$

where Eqs. 8.6 and 8.7 are by virtue of DeMorgan's theorems. Direct implementation of Eq. 8.7 would require 1 5-input *AND* gates, 5 4-input *OR* gates, and 3 *INVERTER*s.

Eq. 8.7 is in what is known as the canonical or standard product of sums form. Each sum term in this equation is known as a maxterm because each contains all four literals. Each maxterm for a row is the complement of the minterm for that row. A short hand way of writing Eq. 8.7 is

$$F = F(A, D, M, S) = \prod M(0, 2, 4, 6, 8) \tag{8.8}$$

where the product symbol indicates a product of sums, the capital M indicates a maxterm, and the numbers in the parentheses are the decimal row number for which F is a logical 0. The maxterms row numbers will always be the complement of the minterm row numbers. Products of sums are rarely used in the implementation of digital logic.

The terms "product" and "sum" that were used above are standard engineering terminology because of the analogous meaning of the operators " + " and " · " from ordinary algebra. It should be borne in mind that the mathematical function that is actually being performed is the logical *AND* and *OR*. It results in less tongue twisting to use the common expressions "product" and "sum".

Neither Eq. 8.3 nor Eq. 8.7 leads to easy implementation; it can be done using these equations but would require a moderately large number of gates. Therefore, before continuing with the design of the control unit for the home security system it will be necessary to examine a method of function minimization. Namely, is it possible to implement the truth table for this system with a function that contains fewer terms and fewer literals? Numerous techniques exist for such function minimization. There are computer algorithms that will automatically determine the function minimization and there are families of programmable logic that will implement the hardware directly from the truth table. The method of function minimization that will be examined in this experiment is known as the Karnaugh map. Karnaugh maps can be constructed for Boolean functions of from 2 to 8 variables. However, when the number of variables exceeds 4 they become cumbersome and unwieldy.

Karnaugh maps are alternative ways of expressing truth tables that were developed by M. Karnaugh in 1953. It consists of arranging the truth table so that simplifications can readily be identified.

8.2.1 Two Variable Karnaugh Map

The Karnaugh map for a function of two variables, $F(X,Y)$, is

		Y 0	1
X	0	m(0)	m(1)
	1	m(2)	m(3)

Two Variable Karnaugh Map

where one variable is listed vertically and the other horizontally. The entries in the truth table are the row numbers of the truth table. If the minterm corresponds to a logical 1, a 1 is entered into the square and if not a 0 is entered.

Example 1

Consider the two variable truth table $F(X,Y)$

NO.	X	Y	F
0	0	0	0
1	0	1	1
2	1	0	1
3	1	1	1

The Karnaugh map for this truth table is given by

		Y 0	1
X	0	0	1
	1	1	1

from which the function in the form of a sum of products is given by

$$F(X,Y) = \overline{X} \cdot Y + X \cdot \overline{Y} + X \cdot Y \tag{8.9}$$

Eq. 8.9 can be simplified as

$$F(X,Y) = X \cdot (Y + \overline{Y}) + \overline{X} \cdot Y = X + \overline{X} \cdot Y \tag{8.10}$$

because $Y + \overline{Y} = 1$. Eq. 8.10 can be simplified as

$$F(X,Y) = X + Y \tag{8.11}$$

by virtue of the observation

$$(X + \overline{X}) \cdot (X + Y) = X \cdot X + X \cdot Y + X \cdot \overline{X} + \overline{X} \cdot Y = X \cdot (Y + 1) + 0 + \overline{X} \cdot Y$$

$$= X + \overline{X} \cdot Y = X + Y \tag{8.12}$$

This simplified, minimized, or reduced relationship

$$F(X,Y) = X + Y \tag{8.13}$$

can be obtained directly from the Karnaugh map representation.

The Karnaugh map discloses that F is 1 for all Y from the second row. It also reveals that F is 1 for all X from the second column. This grouping of 2 1s can be circled as shown. Combining two minterms on a 2 variable Karnaugh map yields a function of one Boolean variable. Thus the sum of products is given by Eq. 8.13. This is not a standard sum of products since there are missing literals; this is highly desirable because it means that the truth table can be implemented with fewer terms.

X \ Y	0	1
0	0	1
1	1	1

8.2.2 Three Variable Karnaugh Map

For a three variable Boolean function the variables and minterms are arranged as shown below. Note that the YZ variable on the top is listed in a Gray code so that only one bit changes when shifting from column to column.

X \ YZ	00	01	11	10
0	m(0)	m(1)	m(3)	m(2)
1	m(4)	m(5)	m(7)	m(6)

Three Variable Karnaugh Map

With the three variable map grouping can be made of groups of 2 or 4 adjacent minterms to reduce the function. A grouping of 4 adjacent minterms produces a product term consisting of 1 literal. A grouping of 2 adjacent minterms produces a product grouping of 2 literals. A minterm which is not adjacent to another represents a product term with all three literals. Always begin by making the largest grouping of minterms (4 then 2)

Example 2

Consider the three variable function

$$F = F(X, Y, Z) = \sum m(0, 2, 5, 6) \tag{8.14}$$

for which the Karnaugh map is given by

X \ YZ	00	01	11	10
0	1	0	0	1
1	0	1	0	1

Example Two

For this map the two minterms $m(2)$ and $m(6)$ are adjacent and can be combined to yield $Y \cdot \overline{Z}$. The minterm $m(5)$ cannot be combined with anything and must be represented as $X \cdot \overline{Y} \cdot Z$. The Karnaugh map is to be considered a topological spherical surface which mean that column 1 is adjacent to column 4 which means that $m(0)$ and $m(2)$ can be combined to yield $\overline{X} \cdot \overline{Z}$. This yields as the function

$$F = F(X,Y,Z) = Y \cdot \overline{Z} + X \cdot \overline{Y} \cdot Z + \overline{X} \cdot \overline{Z} \tag{8.15}$$

for F.

Example 3

Consider the function

$$F = F(X,Y,Z) = \sum m(0,1,2,3,4,6) \tag{8.16}$$

for which the Karnaugh map is

		YZ 00	01	11	10
X	0	1	1	1	1
	1	1	0	0	1

Example Three

For this example the entire top row can be grouped together to yield $\overline{X}$. The four minterms in columns 1 and 4 can be grouped together to yield $\overline{Z}$. So this function is given by

$$F = F(X,Y,Z) = \overline{X} + \overline{Z} \tag{8.17}$$

8.2.3 Four Variable Karnaugh Maps

For a 4 variable function the minterms are arranged on the Karnaugh map as

		YZ 00	01	11	10
	00	m(0)	m(1)	m(3)	m(2)
WX	01	m(4)	m(5)	m(7)	m(6)
	11	m(12)	m(13)	m(15)	m(14)
	10	m(8)	m(9)	m(11)	m(10)

Four Variable Karnaugh Map

With the four variable map, groupings are made for horizontally and vertically adjacent (diagonal doesn't count–this isn't tick-tack-toe) minterms. The number of adjacent minterms must be a power of 2, i.e. 2, 4, 8. One minterm which is not adjacent to anything requires a product terms with 4 literals. A grouping of 2 minterms produces a product term with 3 literals. A grouping of 4 minterms produces a product term with 2 literals. A grouping of 8 minterms produces a product term with 1 literal. The largest grouping of minterms should be made first, and continue until all minterms are included.

Example 4

Consider the function $F = F(W,X,Y,Z)$

$$F = F(W,X,Y,Z) = \sum m(0,1,2,4,6,8,9,10,12,14,15) \tag{8.18}$$

for which the Karnaugh map is given by

WX \ YZ	00	01	11	10
00	1	1	0	1
01	1	0	0	1
11	1	0	1	1
10	1	1	0	1

Example Four

For this problem the entire columns 1 and 4 can be grouped to produce a product term consisting of the single literal $\overline{Z}$. The first two columns of rows 1 and 4 can be grouped to yield a product term with two literals $\overline{X} \cdot \overline{Y}$. Finally, in row 3 the minterms in the third and fourth columns can be combined to yield a product term with three literals $W \cdot X \cdot Y$. Therefore the reduced function is given by

$$F = F(W, X, Y, Z) = \overline{Z} + \overline{X} \cdot \overline{Y} + W \cdot X \cdot Y \tag{8.19}$$

Example 5

Consider the function

$$F = F(W, X, Y, Z) = \sum m(0, 1, 2, 4, 6, 8, 10, 12, 15) \tag{8.20}$$

for which the Karnaugh map is given by

WX \ YZ	00	01	11	10
00	1	1	0	1
01	1	0	0	1
11	1	0	1	0
10	1	0	0	1

Example Five

There are no groups of 8 minterms. The first column has a grouping of 4 minterms which yields the product term with 2 literals $\overline{Y} \cdot \overline{Z}$. Since the map is topologically spherical, the first two rows of columns 1 and 4 form a group of 4 minterms which yields a product term with two literals $\overline{W} \cdot \overline{Z}$. The four corners of the map for a group of 4 minterms which yields a product term with two literals $\overline{X} \cdot \overline{Z}$. The two minterms in the first two columns of row 1 can be combined to yield a product term with three literals $\overline{W} \cdot \overline{X} \cdot \overline{Y}$. Finally, the minterm in row 3 column 3 cannot be combined with another other so it represents the product term with 4 literals $W \cdot X \cdot Y \cdot Z$. Therefore, the function F can be reduced to

$$F = F(W, X, Y, Z) = \overline{Y} \cdot \overline{Z} + \overline{W} \cdot \overline{Z} + \overline{X} \cdot \overline{Z} + \overline{W} \cdot \overline{X} \cdot \overline{Y} + W \cdot X \cdot Y \cdot Z \tag{8.21}$$

Now that the topic of the Karnaugh map function minimization has been examined the design of the control unit for a home security system can be completed. The Karnaugh map for this problem is given by

AD \ MS	00	01	11	10
00	0	1	1	0
01	0	1	1	0
11	1	1	1	1
10	0	1	1	1

Karnaugh Map for Home
Security System

For this problem the second and third column may be combined to yield a product term with a single literal S. The third row can be combined to yield a product term with two literals AD. Finally the 4 minterms in rows 3 and 4 and columns 3 and 4 can be combined to yield a product term with two literals AM. Therefore, the Boolean expression for the alarm for the home security system is given by

$$F = F(A, D, M, S) = A \cdot D + A \cdot M + S = A \cdot (D + M) + S \tag{8.22}$$

Thus the Karnaugh map has yielded what should have been intuitively obvious for the home security system. Namely, the sensor for the smoke detector is one product term in the sum so that if it detects smoke it activates the alarm no matter what else the security system is doing. The arm switch multiplies the sum of the door and motion detector so that if it is off these sensors have no effect on the alarm. Finally, if the system is armed then either the door sensor or the motion detector will activate the alarm. Therefore, Karnaugh maps are an invaluable for design artifice for engineers with poorly developed intuitions.

It may often be the case that for a particular digital logic problem not all 2^N possible combinations are permitted or are relevant. In this case the unspecified output may be marked with an X in the Karnaugh map and be interpreted as either a 1 or 0 at the discretion of the designer. (This is known in the parlance of digital design as a "don't care" or a "wild card"). As an example consider the design of a system to indicate whether a month of the year has an even or odd number of days; it would require 4 binary digits to specify the number 12 but since there are only 12 months in the year the outputs of the system for 1101 to 1111 would be not permitted and could therefore be marked with an X.

8.2.4 Multiplexers

A multiplexer is a combinational logic gate that has 2^N input data lines and one output line; it is also known as a 1 of 2^N data selector (or in the vernacular of ECEs a $2^N \times 1$ MUX). There are N address inputs which determine which of the 2^N input lines are connected to the output. An 2^N input multiplexer can be used to implement a truth table of $N + 1$ Boolean variables.

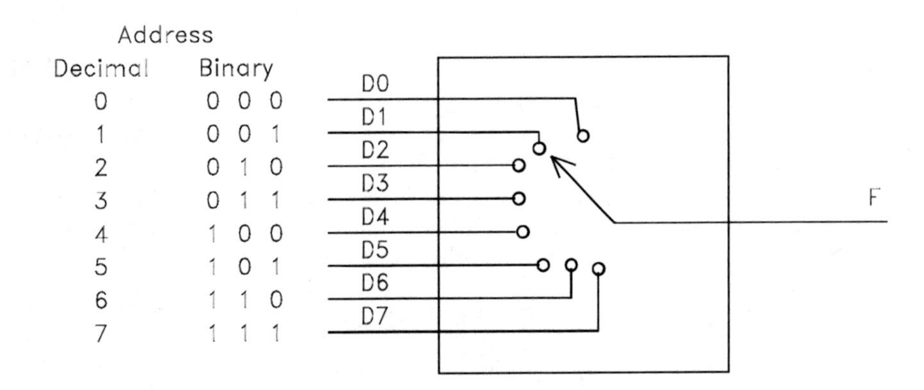

Figure 8.1: An eight input multiplexer.

An eight input multiplexer is shown in Fig. 8.1. It has one output and eight inputs. The three bit binary address determines the position of the switch which routes one of the eight data inputs, D0 — D7, to the output F. The switch is an electronic switch rather than a mechanical one.

It should be abundantly obvious how to implement a truth table with N or less rows with a 2^N to 1 multiplexer. There are 2^N rows in the truth table and each of the 2^N inputs can be made a logic 1 or 0

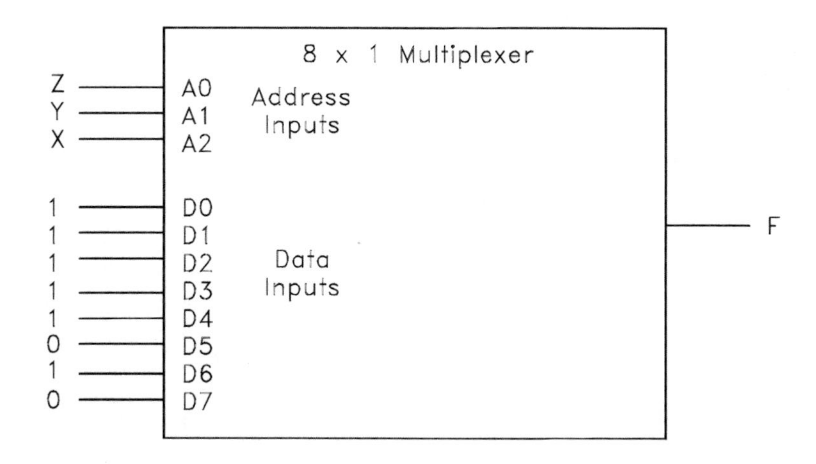

Figure 8.2: Implementation of three variable logical function with an $8x1$ multiplexer.

in accordance to the desired value of F. Shown in Fig. 8.2 is an 8×1 multiplexer which is being used to implement a Boolean function of three variables X, Y, and Z. The three address lines (also called select lines) determine which of the 8 inputs is connected to the output F. A_0 is the LSB and A_2 is the MSB. All 0s selects data input D0 and all 1s selects data input D7.

The data inputs shown would implement the truth table for Example 3. To implement truth tables for other functions of three variables the Ds would have to be change to correspond to that truth table.

The use of a 8×1 multiplexer to implement a function of 4 Boolean variables $[F(W, X, Y, Z)]$ is not quite so obvious as the 3 variable problem. The solution is to make the three most significant binary digits (W, X, and Y) the address lines. The truth table is then split into pairs of two rows. For each pair of rows addressed by the three most significant bits the output is either 0, 1, Z, or $\overline{Z}$. For instance, consider the truth table for the home security system

Row No.	A	D	M	S	F
0	0	0	0	0	0
1	0	0	0	1	1
2	0	0	1	0	0
3	0	0	1	1	1
4	0	1	0	0	0
5	0	1	0	1	1
6	0	1	1	0	0
7	0	1	1	1	1
8	1	0	0	0	0
9	1	0	0	1	1
10	1	0	1	0	1
11	1	0	1	1	1
12	1	1	0	0	1
13	1	1	0	1	1
14	1	1	1	0	1
15	1	1	1	1	1

Truth Table for Home Security System

The variables here are called A, D, M, and S rather than W, X, Y, and Z. The truth table discloses the relationship between F and the least significant bit S for each pair of rows. The implementation of this truth table with the 8×1 multiplexer is then given in Fig. 8.3.

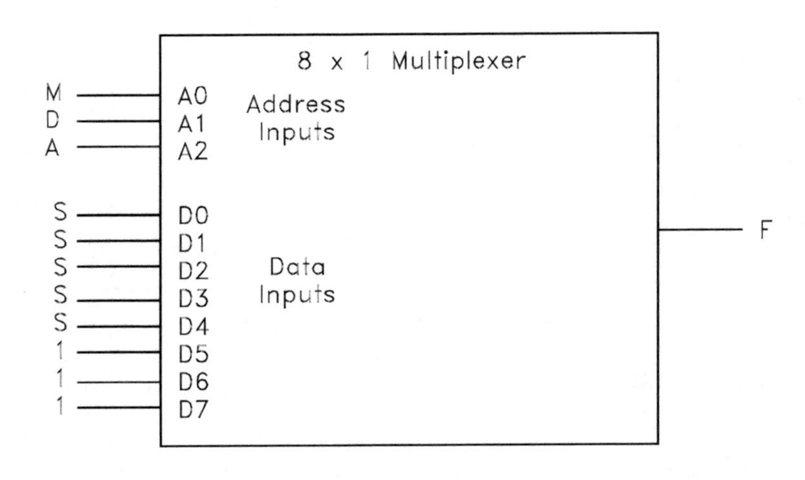

Figure 8.3: Implementation of home security system.

The 8×1 multiplexer that will be used in this experiment is the $74LS151$ shown in Fig. 8.4. This IC has 16 pins whereas the digital IC used in the previous experiments had only 14 pins. Pin 16 must be connected to the $+5\ V$ supply and pin 8 connected to the system ground. This multiplexer has an $\overline{ENABLE}$ input (pin 7) which is active low which means that pin 7 must also be grounded. Pin 5 is the output and pin 6 is the complement of the output.

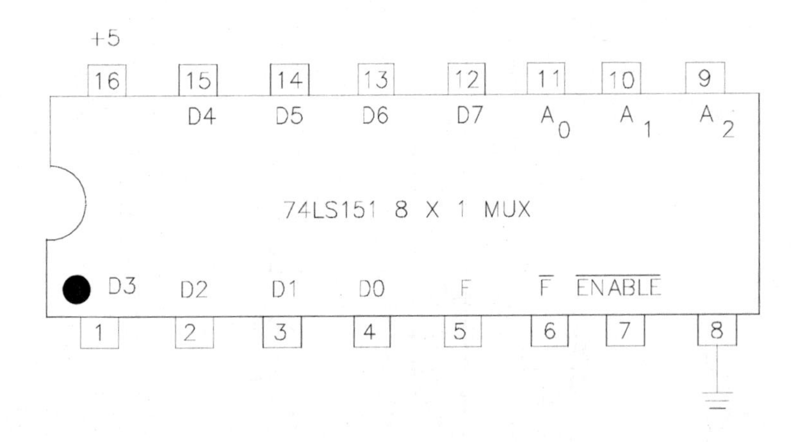

Figure 8.4: The $74LS151$ $8x1$ multiplexer.

8.3 Procedure

8.3.1 Three Variable Function with 1-, 2-, and 3- input gates

The laboratory instructor will specify column F for the Boolean function of three variables shown below, $F(X, Y, Z)$.

Row No.	X	Y	Z	F
0	0	0	0	
1	0	0	1	
2	0	1	0	
3	0	1	1	
4	1	0	0	
5	1	0	1	
6	1	1	0	
7	1	1	1	

$$F = F(X, Y, Z) = \sum m(\cdots)$$

Determine the Karnaugh map for this function.

X \ YZ	00	01	11	10
0				
1				

Three Variable Karnaugh Map

Determine the simplified function from this Karnaugh map.

Sketch a hardware implementation of this simplified function using the 1-, 2-, and 3- input logic gates. Label the pin numbers. (Pinouts for each of these ICs are at the end of the lab manual.)

Assemble the designed circuit with the power off. Remember to connect pin 14 of each of these ICs to the +5 V supply and pin 7 to ground. Use logic switches 1, 2, and 3 on the CADET for X, Y, and Z. Turn the power on.

Verify that the designed circuit reproduces the truth table specified by the laboratory instructor.

Laboratory Instructor Verification ________________

8.3.2 Three Variable Function with Multiplexer

Use the 74LS151 multiplexer to implement the three variable truth table. Draw a diagram of the circuit with the data inputs labeled (corresponding to the row in the truth table). For a data input that is to be a 1, connect it to the +5 V supply and for inputs that are to be 0 connect them to the system ground. Assemble the circuit with the power off. Remember to connect pin 16 of the 74LS151 to the +5 V supply and pin 8 to the system ground. Use the same logic switches assignment that was employed in the previous step.

The pinouts for the 74LS151 are found on page 117 in Fig. 8.4. Note that the $\overline{ENABLE}$ input (pin 7) must also be grounded.

Turn the power on.

Verify that the designed circuit reproduces the truth table specified by the laboratory instructor.

Laboratory Instructor Verification ________________

Four Variable Function with 1-, 2-, and 3- Input Gates

The laboratory instructor will specify column F in the following truth table for the Boolean function of 4 variables, $F(W, X, Y, Z)$.

Row No.	W	X	Y	Z	F
0	0	0	0	0	
1	0	0	0	1	
2	0	0	1	0	
3	0	0	1	1	
4	0	1	0	0	
5	0	1	0	1	
6	0	1	1	0	
7	0	1	1	1	
8	1	0	0	0	
9	1	0	0	1	
10	1	0	1	0	
11	1	0	1	1	
12	1	1	0	0	
13	1	1	0	1	
14	1	1	1	0	
15	1	1	1	1	

$$F = F(W, X, Y, Z) = \sum m(\dots\dots\dots\dots\dots)$$

Determine the Karnaugh map

WX \ YZ	00	01	11	10
00				
01				
11				
10				

Determine the reduced or simplified function from this Karnaugh map.

$$F = F(W, X, Y, Z) =$$

Sketch a hardware implementation of this simplified function using the 1-, 2-, and 3- input logic gates. Label the pin numbers. (Pinouts for each of these ICs are at the end of the lab manual.)

Assemble the designed circuit with the power off. Remember to connect pin 14 of each of these ICs to the +5 V supply and pin 7 to ground. Use logic switches 1, 2, and 3 on the CADET for X, Y, and Z. Turn the power on.

Verify that the designed circuit reproduces the truth table specified by the laboratory instructor.

Laboratory Instructor Verification ______________________________

8.3.3 Four Variable Function with Multiplexer

Implement the four variable truth table using the 74LS151 multiplexer. Remember to connect pin 16 of this IC to the +5 V supply and pin 8 to the ground. Also ground pin 7. Use logic switches 1, 2, 3, and 4 for W, X, Y, and Z. Remember that the data inputs are now either 0, 1, Z, or $\overline{Z}$.

Draw a labeled diagram of the circuit.

Assemble the circuit with the power off. Turn the power on.

Verify that the designed circuit reproduces the truth table specified by the laboratory instructor.

Laboratory Instructor Verification ______________________________

8.4 Laboratory Report

Turn in the sketches that were made for the four circuits that were examined. Answer any supplementary questions that may have been posed by the laboratory instructor.

Chapter 9

Flip Flops and Counters

9.1 Objective

The objective of this experiment is to experimentally examine two basic types of flip flops and their applications in binary counters. The D and JK flip flops will be examined as well as an IC 4 bit counter.

9.2 Theory

Oftentimes a simple knowledge of the inputs to a digital circuit is not sufficient. Consider the case of the car seat belt problem for which an alarm is to sound and the ignition disabled if the driver is seated and the seat belt is not buckled. This problem could be solved with a simple AND gate where the inputs are the sensor for the seat and the seat belt. However, some drivers who dislike seat belts will buckle the seat belt before sitting down and defeat this simple combinational logic problem. This requires that a circuit be designed that will enable the ignition and disable the alarm if and only if the driver sat down before buckling the seat belt. This requires a type of digital circuit known as a sequential circuit.

Flip flops are binary storage elements capable of storing one bit of information. They are one of the most fundamental elements of digital sequential circuits. Sequential circuits are circuits where the time order of the digital signals is of importance in contrast to combinational circuits where time was not a variable. An asynchronous sequential digital circuit is one where the output is a function of the order in which the inputs are applied. A synchronous sequential digital circuit contains a clock signal which is a periodic function of time so that the occurrence of certain events can occur at only certain time points on the clock signal. The clock signal is analogous to the speed at which a conveyor belt moves on an assembly line in a factory. Flip flops differ according to the number of inputs and outputs. Only two types of flip flops will be considered in this experiment: the D and the JK.

The D or delay type flip flop to be considered in this experiment is shown in Fig. 9.1 along with its function table. There are different types of D flip flops but the function table shown in Fig. 9.1 is for the TTL 74LS74 D flip flop that will be used in this experiment. It has two outputs Q and $\bar{Q}$ which are normally complements of each other. When $Q = 0$ and $\bar{Q} = 1$ the flip flop is said to be reset (R) and when $Q = 1$ and $\bar{Q} = 0$ it is said to be set (S). It contains four inputs D, C, S, and R. The C input is the clock signal and D is the data input. The D input is known as a synchronous input since its effect on the state of the flip flop depends on the clock input. The other two inputs are S (set) and R (reset) which are known as asynchronous inputs since their effect on the state of the flip flop is independent of the clock input.

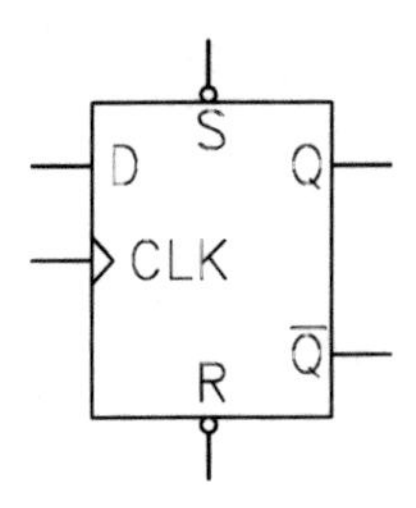

Figure 9.1: D flip flop.

Function Table

Inputs				Outputs	
S	R	C	D	Q	$\overline{Q}$
L	H	X	X	1	0
H	L	X	X	0	1
L	L	X	X	1	1
H	H	↑	H	1	0
H	H	↑	L	0	1

The function table for the D flip flop gives the output corresponding to the various possible inputs. If S is L (low corresponding to a logical $0 \cong 0$ Volts) and R is H (high corresponding to a logical $1 \cong +5\ V$), then the flip flop is set with $Q = 1$ and $\bar{Q} = 0$ no matter what the state of the other two inputs (D and C). (The X in the function table is known as a don't care which means that the variable can be either 0 or 1.) Conversely, if S is H and R is L, then the flip flop is reset with $Q = 0$ and $\bar{Q} = 1$. If both S and R are low then both outputs are one; this condition is never (well hardly ever) used because it places the flip flop in an indeterminate state. The bubble or circle on the S and R inputs means that the active input is low.

If both S and R are inactive (H), then the output is determined by the synchronous inputs D and C. The arrow in the function table means that when the clock signal goes high (from $0\ V$ to $+5\ V$) that the Q output becomes equal to the D input at the positive transition of the clock. This type of flip flop is known as an edge triggered flip flop because the action of the synchronous inputs occurs on the transition of the clock. Thus the Q output is the D input delayed by one clock cycle.

Function Table

Inputs		Outputs
J	K	Q_{N+1}
0	0	Q_N
0	1	0
1	0	1
1	1	$\overline{Q_N}$

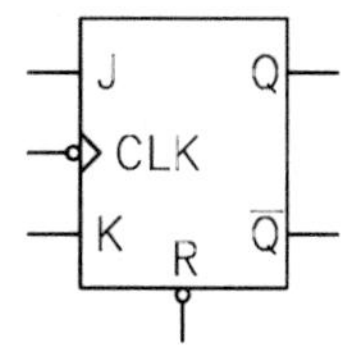

Figure 9.2: JK flip flop.

The JK flip flop of the type to be used in this circuit is shown in Fig. 9.2. It has two outputs Q and $\bar{Q}$ which can never be the same, one asynchronous input R, and two synchronous inputs J and K and the clock C. There are other types of flip flops but the one shown in Fig. 9.2 corresponds to the TTL 74LS107 that will be used in this experiment.

The R input of the JK flip flop is used to reset the flip flop. When the R input is low the $Q = 0$ and $\bar{Q} = 1$. The bubble by the R input means that it is active low.

This type of flip flop (74LS107) is known as a master slave flip flop. It contains two circuits known as latches. The master latch accepts data when the clock signal goes high and the slave latch outputs this data when the clock signal goes low. The bubble by the clock input means that the outputs change state when the clock signal goes low.

If the R input is high, the outputs are determined by the synchronous inputs J, K, and C. If both J and K are both low or logical zeros then the output of the flip flop does not change or retains its current state after the application of the clock signal. If J is high and K is low, then the flip flop sets after the application of the clock. If J is low and K is high, then the flip flop resets after the application of the clock input. If both J and K are high, the flip flop changes to the complement of its current state or toggles after the application of the clock signal.

One of the most important applications of flip flops are counters. In the most general sense, counters are digital sequential circuits that cycle through a predetermined set of states in accordance with a clock signal. The two types of counters that will be considered in this experiment are natural binary counters and divide by N counters.

Natural binary counters are designed to count the number of pulses applied to the input of the counter; they either count up or down in the binary number system. Between pulses the binary count is stored in the flip flops. An N bit counter would use N flip flops and would be capable of storing a count from all 0's (N $0's$) to all 1's (N $1's$ which equals the decimal number $N - 1$). Such a counter is said to have N stages. All N bit binary counters are divide by 2^N counters but it is possible to implement a divide by 2^N counter by using fewer than N flip flops.

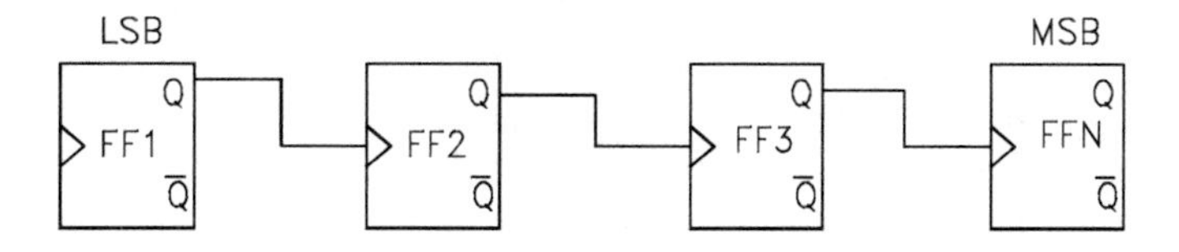

Figure 9.3: Ripple counter.

A common method to implement a binary counter with N flip flops is shown in Fig. 9.3. Initially all the flip flops are reset so that the Q outputs are all 0. Each flip flop is set to toggle or change states when a clock pulse is applied to the clock input and the Q output of the flip flop is connected to the clock input of the next flip flop. Thus the count ripples down the N flip flops and this type of counter is known as a ripple counter. The least significant bit (LSB) is the state of the flip flop on the far left and the most significant bit (MSB) is the state of the flip flop on the right. This type of counter is known as a binary ripple up counter since it is used to count from a binary number consisting of all 0's to the binary number consisting of all 1's. Either D or JK flip flops can be used to implement these ripple counters.

If the $\bar{Q}$ outputs are connected to the clock input of each succeeding stage, a ripple down counter is obtained. This counter would initially have all the flip flops set so that the Q output of each would initially be 1. Each time a clock pulse is applied to the input the count of the counter decreases by 1.

For large counters ripple counters have the disadvantage that a long time is required for an input pulse to the LSB flip flop to propagate down the counter to the MSB flip flop. Therefore, when N is large, synchronous counters are used. These counters have the same clock signal applied to the clock input of each of the N flip flops and combinational logic is used to obtain either the D or JK inputs required to implement an up or down counter. With this type of counter all stages change simultaneously in response to the common clock input.

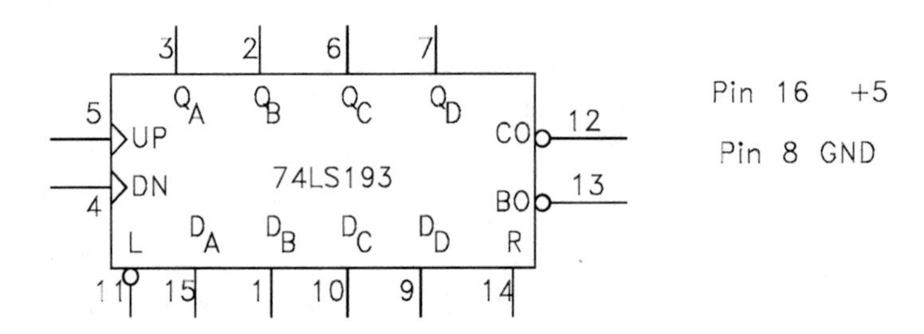

Figure 9.4: 74LS193 four bit binary $UP/DOWN$ counter.

Counters are such fundamental elements of digital systems that the flip flops are often integrated into a single package that is intended only for use as a counter. The 74LS193 4 bit binary up/down counter is shown in Fig. 9.4. This counter contains four flip flops and can be used to counter either up or down. When the load input is made low the counter can be set to the initial state determined by the four D inputs where D_A is the LSB and D_D is the MSB. The output of this counter are the four Q outputs where Q_A is the LSB and Q_D is the MSB. The counter can be reset (all four $Q's$ 0) by making the R input high. Each time a positive pulse is applied to the count up input (UP) the counter counts up one count and each time a positive pulse is applied to the count down input (DN) the counter decrements by one.

If the 74LS193 is counting up and the count reaches four 1's the next pulse applied to the count up input will cause the counter to overflow. This resets the counter to all 0's and causes the count out output to go low. Conversely, when the counter is counting down and reaches a count of all 0's the next pulse applied to the count down input causes the counter to underflow which resets the Q to all 1's and causes the borrow out (BO) output to go low. Thus any number of these counters could be cascaded to provide an N bit binary up/down counter.

Most humans prefer a digital count to a binary count because the digital number system in imbedded in our culture. Therefore, specials ICs have been developed to convert binary counts into digital counts. These involve a special IC known as a seven segment display because 7 LEDs are used to represent the 9 decimal digits. The output of the counter is connected to an IC known as a display driver (Fig. 9.5) or binary coded decimal to 7-segment display decoder/driver. A common cathode 7-segment display of the type that will be used in this experiment is shown in Fig. 9.6 (the actual pinouts for the IC used in the experiment may differ from this because this is not standardized.). It is called a common cathode device because the 7 LEDs have their cathodes connected internally. The seven segments are labeled "a" through "g".

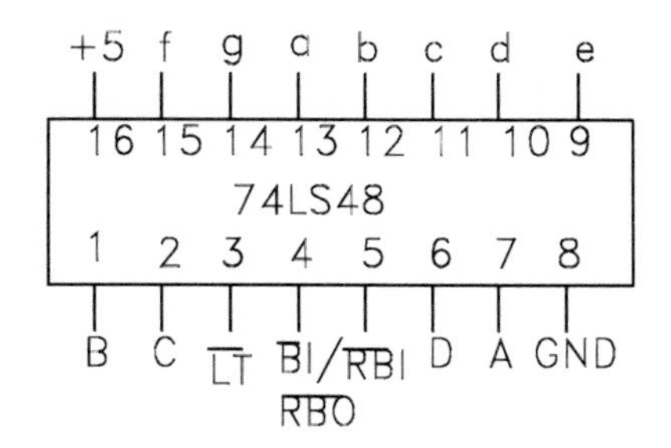

Figure 9.5: 74LS48 BCD to 7-segment display driver.

The 7-segment display cannot be driven directly from the counter output ($Q_D Q_C Q_B Q_A$). Another IC known as a 74LS48 BCD to 7-segment display driver or decoder will be used. This IC is shown in Fig. 9.5. The four inputs are labeled "A" through "D". The seven output segments are labeled "a" through "g". The input labeled "LT" is known as the "lamp test" which, if made low, will cause all 7 segments to be illuminated simultaneously. The inputs labeled "$\overline{BI}/\overline{RBO}$" and "$\overline{RBI}$" are used when counters along with their displays are to be cascaded.

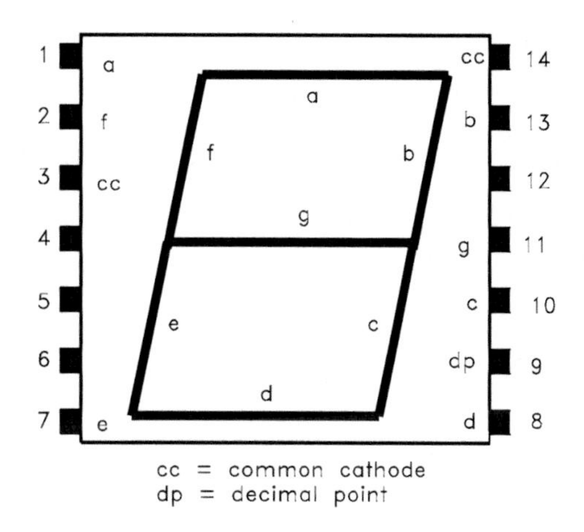

Figure 9.6: Common cathode 7 segment display.

9.3 Procedure

9.3.1 Debounced Switch

This experiment will use sequential logic elements which require an input known as a "debounced input". Ordinary switches, such as the LS1 through LS8 switches on the **CADET** that are used for the other inputs, do not close cleanly, i.e. they bounce or make and break connects many times before finally coming to rest when switched from one position to another. This would be unacceptable for a clock input to a flip-flop or counter since each of the bounces would be interpreted as a clock pulse. Therefore, for the clock input to a sequential logic element a "debounced switch" is required which makes or breaks the connection only once. This debounced switch is an electronic switch. Two such switches are available on the **CADET**, PB1 and PB2 (push buttons one and two).

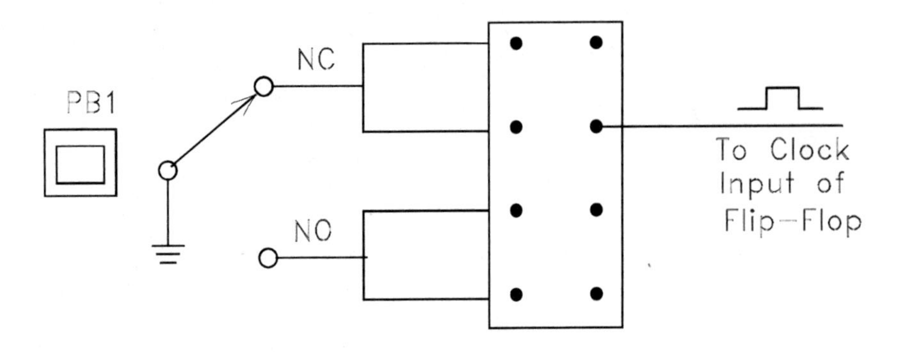

Figure 9.7: Debounced switch for positive pulse.

Any of the 4 points by NC (normally closed) on the breadboard socket shown in Fig. 9.7 may be used for a positive pulse. (A positive pulse is one which goes from 0 to + 5 Volts when the button is pushed.) The four points attached to NO (normally open) are used for negative pulses.

9.3.2 Divide by 2 Counter

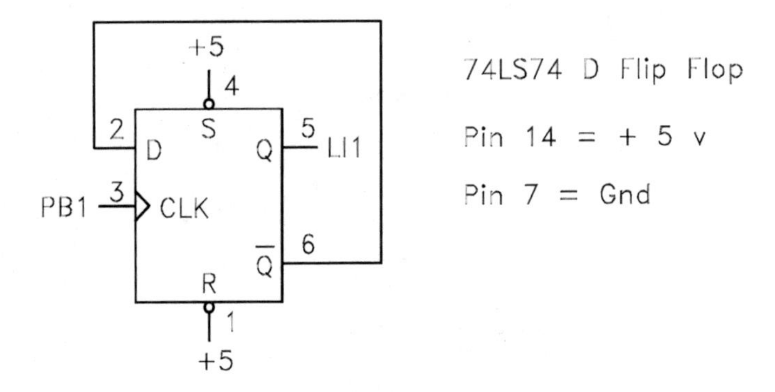

Figure 9.8: Divide by 2 counter.

Assemble the circuit shown in Fig. 9.8 with the power off. The logic indicators LIs are located at the right on the **CADET**. Turn the power on. Push the PB1 button successively and note the effect on the Q output. How many times must the PB1 button be pressed before the Q output goes through one complete cycle?

Remove the PB1 connection from pin 3 and connect the TTL clock to pin 3. Set the frequency to 1 Hz and note and describe the display.

Remove the connection to the logic indicator. Change the clock frequency to 1 kHz. Turn on the oscilloscope and permit it to warm up for around 1 minute. Press **Default Setup**. Connect the $CH1$ input to the TTL clock through a 2.2 kΩ (Place a 2.2 kΩ resistor in series between the clock or TTL output and the $CH1$ oscilloscope lead.) and $CH2$ to the Q output. (Use the BNC connectors on the bottom of the **CADET** to make this connection.) Press **Autoset** on the oscilloscope. Turn Ch2 on. Turn on the measurements to determine the frequency of the signals on the two channels. (**Measure, Add Measurement, Frequency Ch1, Frequency Ch2**) If a stable display cannot be obtained, press **Run/Stop** or **Single** on the oscilloscope. Print the waveforms that appear. Demonstrate the properly functioning circuit to the laboratory instructor.

Laboratory Instructor Verification ____________________

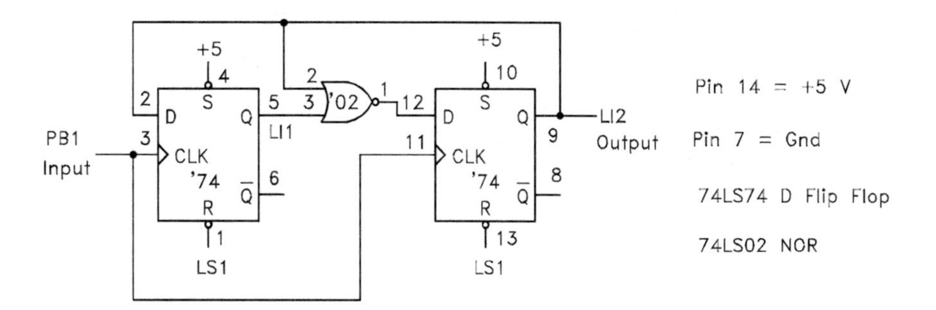

Figure 9.9: Divide by 3 counter.

9.3.3 Divide by 3 Counter

Assemble the circuit shown in Fig. 9.9 with the power off. Two digital ICs are used in this circuit: 74LS74 and 74LS02 (Since the 74LS prefix is common to all the digital ICs used, it is replaced in the figures with an apostrophe.) Connect pin 14 of each IC to the +5 V bus and pin 7 of each to ground. The logic switch LS1 is located at the bottom of the **CADET.** Turn the power on. Set logic switch 1 (LS1) to 0 (low) and then 1 (high); this resets the Q outputs to 0. Press the PB1 button and note the effect on the Q output of the flip flop on the right (pin 9).

Disconnect the PB1 button from the clock inputs and replace it with the TTL clock signal. Set the frequency to 1 Hz and describe the display.

Remove all connections to the logic indicators. Change the frequency of the clock to 1 kHz and connect the clock input to the $CH1$ and the output to $CH2$ of the oscilloscope. Press **Autoset**, rotate the Horizontal

Scale knob until the time per division is 1 ms, press **Run/Stop** or **Single**, and print the display. (Place a 2.2 kΩ resistor in series between the clock or TTL output and the $CH1$ oscilloscope lead.) Demonstrate the properly functioning circuit to the laboratory instructor.

Laboratory Instructor Verification ______________________________

9.3.4 4 Bit Binary Counter

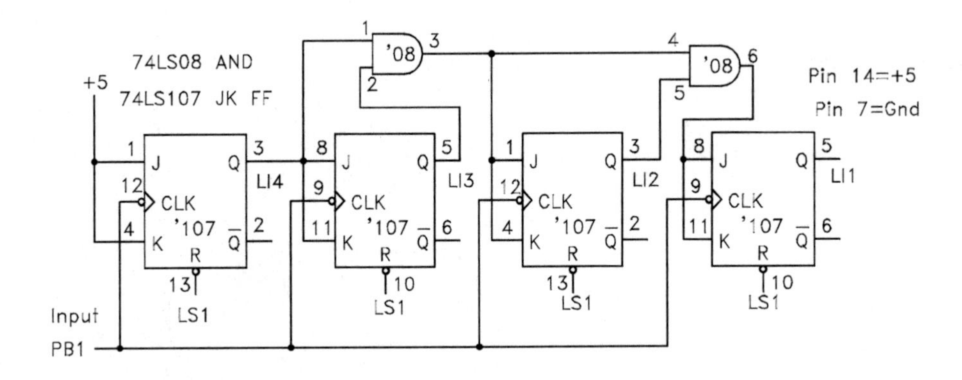

Figure 9.10: 4 bit binary counter.

Assemble the circuit shown in Fig. 9.10 with the power off. Two 74LS107s are required for this circuit. The pin numbers for the two leftmost flip flops are for one IC package and the pin numbers for the other two flip flops are for the other package. A 74LS08 AND gate is also required. Connect pin 14 of all ICs to the +5 V supply and pin 7 of each to ground. Turn the power on. Set logic switch 1 (LS1) to 0 and then 1; this resets the four Q outputs to 0. Successively push PB1 and record the count.

Clock Pulse	LI1	LI2	LI3	LI4
0				
1				
2				
3				
4				
5				
6				
7				
8				
9				
10				
11				
12				
13				
14				
15				

Replace the PB1 connection with the TTL clock with its frequency set to 1 Hz. Observe and describe the output display.

Remove all connections to the logic indicators. Change the frequency to 1 kΩ and connect $CH1$ of the oscilloscope to the TTL clock and $CH2$ to the Q output of the final flip flop. (Place a 2.2 kΩ resistor in series between the clock or TTL output and the $CH1$ oscilloscope lead.) Press *AUTOSET*, change the *HORIZONTAL SCALE* so that the time per division is 4 ms, press either **Run/Stop** or **Single**, and print the resulting display. Demonstrate the properly functioning circuit to the laboratory instructor.

Laboratory Instructor Verification ______________________

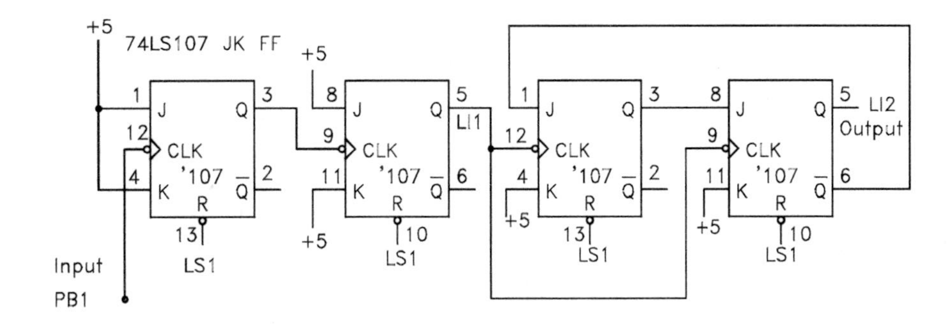

Figure 9.11: Divide by 12 counter.

9.3.5 Divide by 12 Counter

Assemble the circuit shown in Fig. 9.11 with the power off. Turn the power on. Set LS1 to 0 and then 1; this reset each of the four flip flops. Successively push PB1 until the output is something other than 0. How many pushes does this require of PB1?

Remove the PB1 connection to the circuit and replace it with the TTL clock. Set the clock frequency to 10 Hz and describe the display.

Remove all connections to the logic indicators. Change the frequency to 1 kHz and connect $CH1$ of the oscilloscope to the TTL clock and $CH2$ to the Q output of the final flip flop. (Place a 2.2 kΩ resistor in series between the clock or TTL output and the $CH1$ oscilloscope lead.) Press **Autoset**, change the *HORIZONTAL SCALE* so that the time per division is 4 ms, press either **Run/Stop** or **Single**, and print the resulting display. Demonstrate the properly functioning circuit to the laboratory instructor.

Laboratory Instructor Verification ______________________________

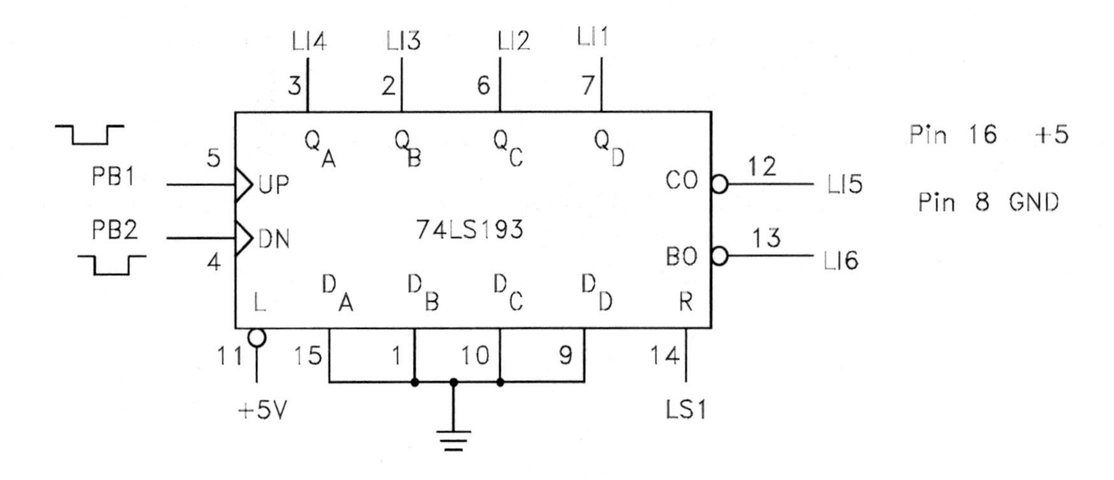

Figure 9.12: 4 bit binary UP/DOWN counter.

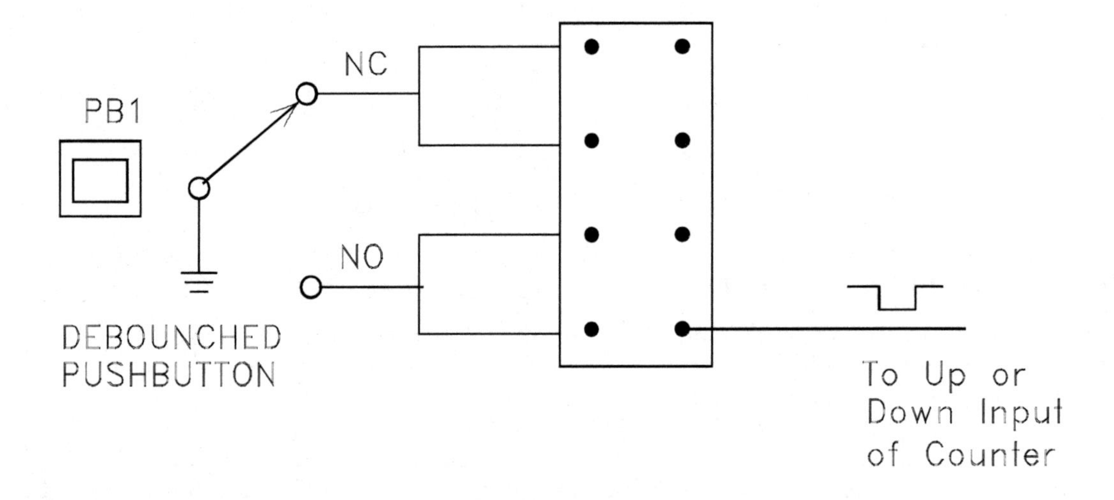

Figure 9.13: Debounced switch for negative pulse.

9.3.6 4 Bit Binary $UP/DOWN$ Counter

Assemble the circuit shown in Fig. 9.12 with the power off. Note that this IC has 16 pins rather than 14. Connect pin 16 to the +5 V supply and pin 8 to ground. Note that the pin locations shown in this figure do not correspond to the physical location of the pins but are arranged according to the logical function. Since two "down pulses" are required they may be obtained from PB1 and PB2 using the four bottom pins as shown in Fig. 9.13.

Turn the power on. Set logic switch 1 (LS1) to 1 and then 0; this resets the 4 Q outputs. Pin 5 is the count up pin and pin 4 is the count down pin. Successively press the PB1 input and obtain the outputs for each successive clock pulse until the counter cycles or repeats. Repeat for PB2. Record the state of CO (logic indicator LI5) when Q_A, Q_B, Q_C, and Q_D are all 1 and PB1 is depressed and momentarily held down. Record the state of BO (logic indicator LI6) when the outputs are all 0 and PB2 is depressed and momentarily held down.

Count Up Function Table

Clock	LI1	LI2	LI3	LI4	LI5	LI6
0						
1						
2						
3						
4						
5						
6						
7						
8						
9						
10						
11						
12						
13						
14						
15						

Count Down Function Table

Clock	LI1	LI2	LI3	LI4	LI5	LI6
0						
1						
2						
3						
4						
5						
6						
7						
8						
9						
10						
11						
12						
13						
14						
15						

Laboratory Instructor Verification ________________________

9.3.7 Seven Segment Display

Assemble the circuit shown in Fig. 9.14 with the power off. Note that the 74LS48 BCD to Seven Segment display decoder, and the 74LS193 4 bit counter have 16 pins whereas the 74LS08 AND, and the 74LS32 OR have 14 pins. Turn the power on. Set LS1 to 1 and then to 0 which resets the counter. Successively press PB1 and note the effect on the seven segment display.

____

Press the PB1 input until the display is 9. The press PB2 and note the effect on the display.

____

Demonstrate the properly functioning circuit to the laboratory instructor.

Laboratory Instructor Verification ________________________

Remove the PB1 connection and replace it with the TTL clock set to 1 Hz. Observe the display. Change the frequency to 1 kHz and remove the clock lead. The display should stop at a decimal digit; this is the basis of a game known as electronic dice.

Turn off the **CADET** and remove the ICs and hook-up wire.

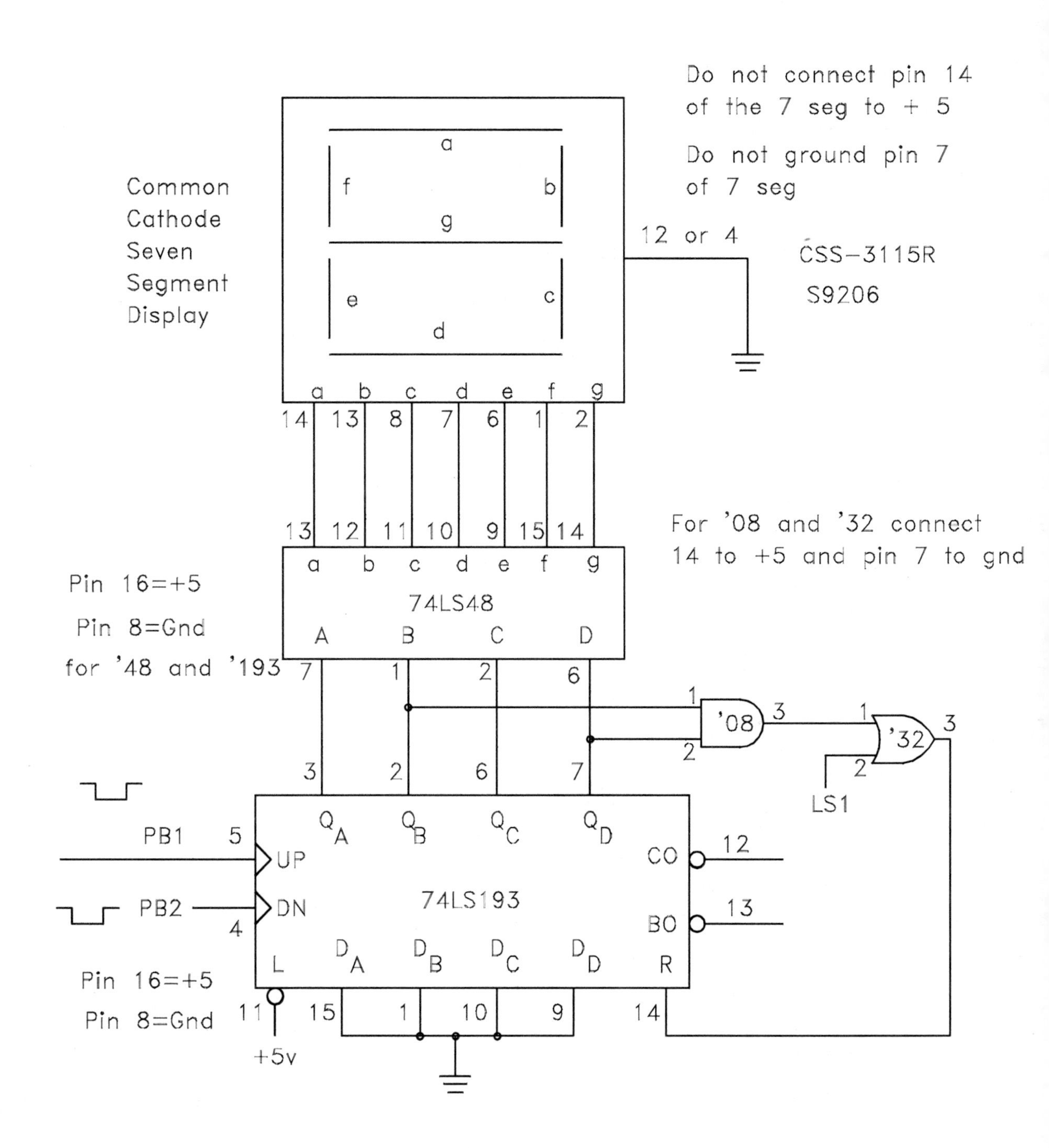

Figure 9.14: Seven segment display.

9.4 Laboratory Report

1. For the 74LS74 D flip flop why is the $S = R = 0$ input known as a disallowed state?

2. Is the divide by 3 counter considered in step 9.3.3 a ripple or synchronous counter? Explain:

3. Is the divide by 12 counter considered in step 9.3.5 a ripple or synchronous counter? Explain:

4. Explain the purpose of the AND gate in the circuit in Fig. 9.14? (Hint: How many fingers and toes does the standard humanoid biped have?)

Turn in all sketches, experimentally obtained function tables, the answers to the questions as well as any supplementary questions that may have been posed by the laboratory instructor.

Chapter 10

Shift Registers and Pseudo-Random Sequences

10.1 Objective

The objectives of this experiment are to experimentally examine the basic properties of shift registers and their application in producing pseudo-random sequences. A four bit shift-register will be implemented using four D type flip flops and an eight bit IC shift register will examined.

10.2 Theory

A register is a digital storage device. An N stage resister is constructed by using N flip flops and can therefore store N bits of information. A special type of register known as a shift register will be the subject of this experiment. Registers are integral components of digital systems and particularly computers.

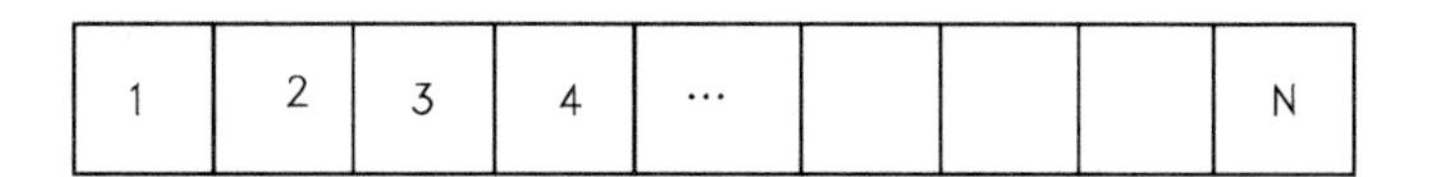

Figure 10.1: An N stage shift register.

Shown in Fig. 10.1 is an N stage shift register. Each stage of this shift register simultaneously shifts its data to an adjacent flip flop or stage in the register upon the application of a clock pulse. Data can be shifted either to the flip flop on the right or the left depending upon the design of the shift register but, obviously, each stage must shift its data in the same direction. If data is shifted to the right it is known as a shift-right register and conversely if data is shifted to the left it is known as a shift-left shift register. Clock pulses are applied synchronously to each of the flip-flops in the N stages so that they all shift together.

If the shift register shown in Fig. 10.1 is assumed to be a shift-right shift register, then after the application of a clock pulse the contents of the Nth stage leave the shift register and the contents of the first stage must come from an external circuit. Where the output data goes and where the input data comes from depends upon the application intended for the shift register. Sometimes the output is connected directly to the input so that after the application of N clock pulses the contents of the register are unchanged. Such registers are often used in arithmetic operations such as the addition, multiplication, masking, and cross correlation of binary numbers.

The most common applications of shift registers are to perform serial-to-parallel conversions and the complement of this operation, a parallel-to-serial conversion. Data is usually processed by a computer is parallel form because it is much faster. However, it is cheaper to obtain data from a peripheral device in serial form because it requires only one data line instead of N data lines. To process the serial data coming into a computer in parallel form a serial-to-parallel converter is required. Conversely, if data is to be transferred in serial form from a computer to a peripheral device a parallel-to-serial converter is required.

The principle application of a shift register that will be examined in this experiment is to produce a pseudo-random sequence. To understand what a pseudo-random sequence is one must first understand what a random sequence is.

A random sequence of binary digits is a sequence of ones and zeros that is completely unpredictable or random. This is what would be obtained if one tossed a fair coin and recorded a head as a binary one and a tail as a binary zero. The probability that the outcome of a coin toss would be a head or a tail is 50 % for either outcome if the coin is fair. A knowledge of the outcome of all previous tosses of this coin would provide no knowledge about the next toss of the coin. Thus, a random sequence of binary digits is what a binary random number generator would produce.

A pseudo-random sequence is one that appears to be random for K bits and then repeats itself. For any sequence of bits less than K, however, it appears to be perfectly random. If K can be made large, then this pseudo-random sequence can be considered to be a random sequence of binary digits.

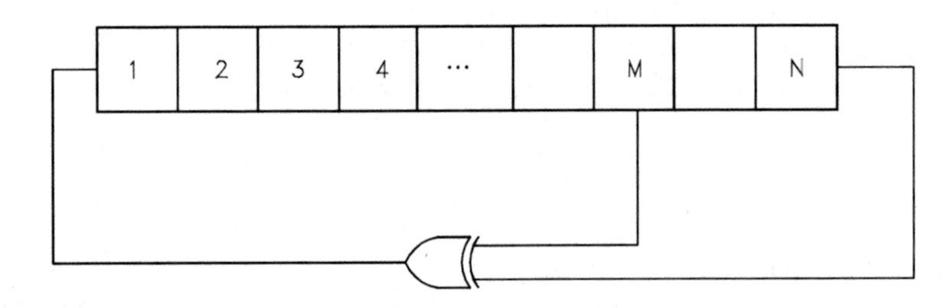

Figure 10.2: A pseudo-random sequence generator.

An electronic circuit that can be used to produce a pseudo-random sequence is shown in Fig. 10.2. It uses an N stage shift register and an *EXCLUSIVE OR* gate. The inputs to the *EXCLUSIVE OR* gate are the output of the Nth and Mth stage of the register. Each time a clock pulse is applied data shifts one stage to the right. The input to the first stage is the *EXCLUSIVE OR* of the Nth and Mth stages and the contents of the Nth stage exit the register. If M and N are properly chosen, then the period of the pseudo-random sequence is given by

$$K = 2^N - 1 \tag{10.1}$$

which makes the period of the signal at the output of the shift register

$$T_s = KT_c = (2^N - 1)T_c \tag{10.2}$$

where T_c is the period of the clock signal. Arbitrary values of M and N will result in values of K that are less than that given in Eq. 10.1. Once N is selected, a specific value of M must be chosen to make K the expression given in Eq. 10.1; some choices of N may require the *EXCLUSIVE OR* or more than the N th and M th stage. If the feedback stages are chosen so that Eq. 10.1 is valid, then the pseudo-random sequence produced by this shift register is known as a maximal length shift register sequence. The selection of M and N to produce a maximal length pseudo-random sequence is a subject of a branch of higher mathematics known as linear algebra and involves determining whether certain polynomials are irreducible and prime over a Galois field. (Galois was a French mathematician who died in a duel over a coquette when he was 20 years old. Although he desperately tried to withdraw from the duel, the cavalier he insulted refused to accept his apology. The night before the duel he wrote his theory of mathematics which was published posthumously; it was so advanced that none of his contemporaries even understood it. Indeed, he failed the entrance exam of a university because he failed to provide his professors with intermediate results. His failure was assured when he became so enraged with the examiner that he struck him in the face with an eraser.)

Pseudo-random sequences are also known as pseudo-noise (PN), pseudo-random generators (PRG), or simply shift register sequences or digital noise. It is the exponential dependence of the period of the sequence on the length of the register that makes it useful for producing long or essentially random sequences. For instance, if the clock frequency is chosen as 33 Mhz and N as 24, then the period of the pseudo-random sequence is only 0.508 seconds. However, if N is increased to 56 then the period becomes 69.24 years. Increasing N to 72 increases the period to 4.54 million years which according to current anthropological theories is about the amount of time that man has been on the planet. Finally, if N were made 96 it would increase the period to 7.613×10^{13} years which is approximately 5,000 times longer than the estimated age of the universe which is 15 billion years by current cosmological theories. Therefore, it isn't difficult to obtain sequence periods that are long even with relatively high clock frequencies such as 33 MHz.

Pseudo-random sequences have numerous applications. They are used to scramble data for security reasons. Data to be scrambled is *EXCLUSIVE OR*ed with a certain pseudo-random sequence known only to the valid users of the data. The data can then be unscrambled by *EXCLUSIVE OR*ing the scrambled data with the same pseudo-random sequence. They are also used in radar ranging due to their sharp autocorrelation properties and to produce a certain type of modulation known as spread spectrum. Spread spectrum is used in both cordless and cellular phones to eliminate noise, interference, and to prevent eavesdropping.

If a pseudo-random sequence is applied to the input of a low-pass filter, the output is white bandlimited Gaussian noise. Thus these sequences are used to produce analog noise. Analog noise is used as a test signal for circuits along with sine and square waves. White noise applied to a speaker has a soothing effect on most people and is used as a sound effect to induce sleep and relaxation.

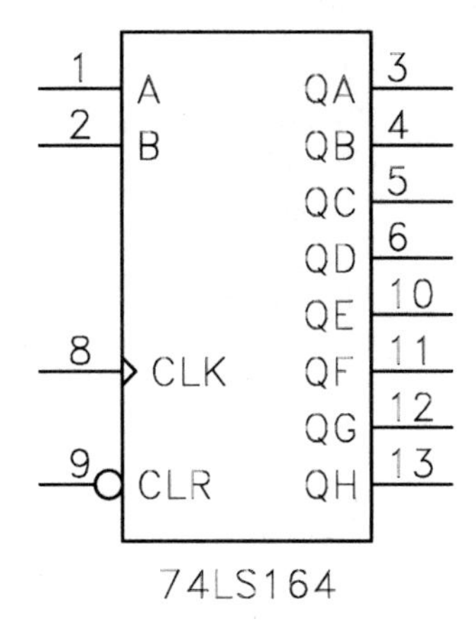

Figure 10.3: 74LS164

Shown in Fig. 10.3 is the 74LS164 TTL 8 bit shift register. It is an 8 bit serial to parallel converter. There are two serial inputs (pins 1 and 2); if there is to be only one serial input these two pins are usually connected together. The 8 parallel outputs are Q_A through Q_H. Each time a clock pulse is applied the contents of the serial input is transferred to Q_A, Q_A to Q_B, Q_B to Q_C, Q_C to Q_D, Q_D to Q_E, Q_E to Q_F, Q_F to Q_G, and Q_G to Q_H and the contents of Q_H exit from the shift register. There is an asynchronous reset input that is active low so that if this input is made low all of the parallel outputs are reset to low. The primary application of this IC is to convert serial data into 8 bit binary words. They can be cascaded so that longer parallel words can be implemented.

10.3 Procedure

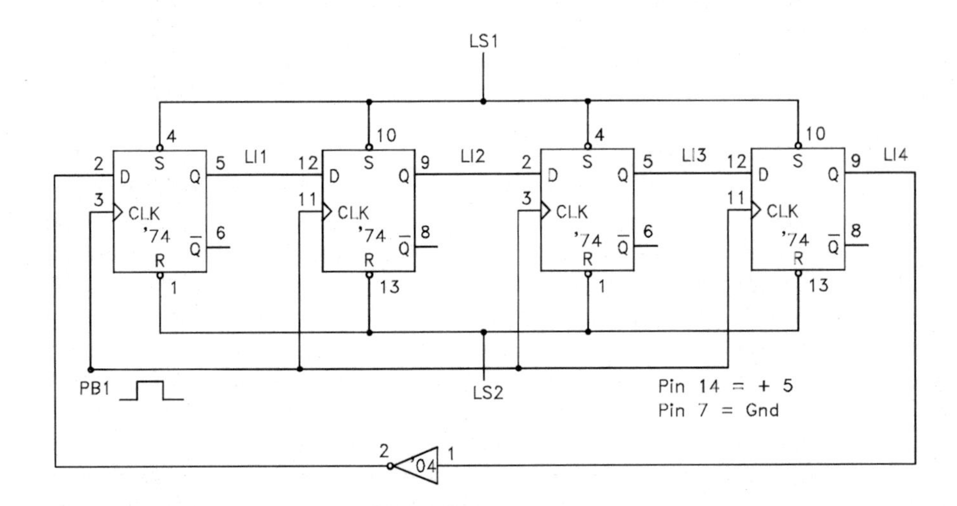

Figure 10.4: 4 bit shift register.

10.3.1 Four Bit Shift Register

Assemble the circuit shown in Fig. 10.4 with the power off. This will require two 74LS74 *D* type flip flop ICs and one 74LS04 *INVERTER*. Begin by establishing busses for +5 V and ground. Connect pin 14 of each of the three ICs to +5 V and pin 7 of each to ground.

Turn the power on. Set both logic switches one (LS1) and two (LS2) to +5 V. Switch logic switch 2 to 0 and then back to 1; this resets all four of the flip flops. Successively press the PB1 button and note and describe the effect on the four logic indicators.

Switch logic switch 1 to 0 and then back to one. Repeat the above procedure.

Demonstrate the properly functioning circuit to the laboratory instructor.

Laboratory Instructor Verification ____________________

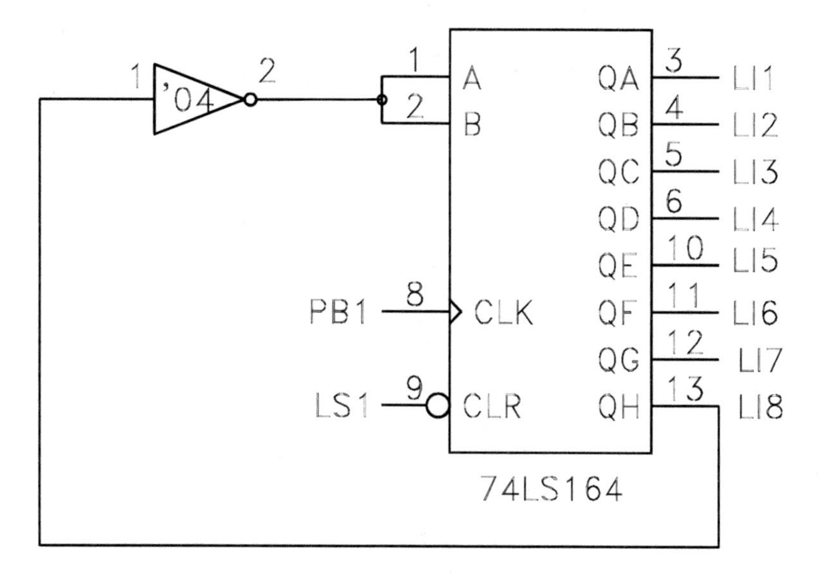

Figure 10.5: 8 bit shift register.

10.3.2 Eight Bit Shift Register

Assemble the circuit shown in Fig. 10.5 with the power off. This will require one 74LS164 eight bit shift register and one 74LS04 *INVERTER*. Begin by connecting pin 14 of each IC to the +5 V supply and pin 7 of each to ground.

Turn the power on. Switch logic switch 1 to the 0 position and then to the one position; this will reset each of the 8 outputs to low or zero. Successively press PB1 and note the effect on the display.

___

___

___

Demonstrate the properly functioning circuit to the laboratory instructor.

Laboratory Instructor Verification ___

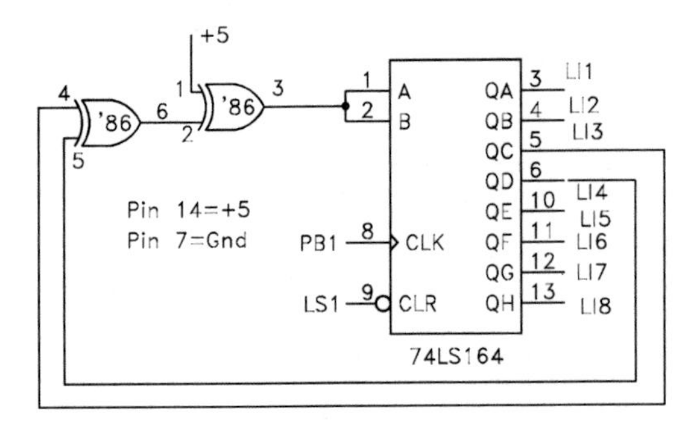

Figure 10.6: 4 stage pseudo-random sequence generator.

10.3.3 4 Stage Pseudo-Random Sequence Generator

Assemble the circuit shown in Fig. 10.6 with the power off. This will require one 74LS164 8 bit shift register and one 74LS86 *EXCLUSIVE OR* gate. Begin by connecting pin 14 of each IC to the +5 V supply and pin 7 to the ground. Since this circuit will use only the first 4 stages of the shift register with the feedback coming from the third and fourth stage, this will produce a pseudo-random sequence with a period of $K = 2^N - 1 = 2^4 - 1 = 15$.

Turn the power on. Set logic switch one to zero and then back to one; this will reset all eight outputs to 0. Successively press PB1 and record the outputs.

Clock Pulse	LI1	LI2	LI3	LI4	LI5	LI6	LI7	LI8
1								
2								
3								
4								
5								
6								
7								
8								
9								
10								
11								
12								
13								
14								
15								
16								

Remove the PB1 input from pin 8 of the 74LS164 and connect the TTL clock to this pin. Set the clock frequency to 1 Hz. Set LS1 to 0 and then to one and time (with a wrist watch) how long it takes for the first four logic state indicators to go through one cycle.

Turn on the oscilloscope. Allow a few minutes for it to warm up. Connect $CH1$ of the oscilloscope to the clock input of the 74LS164 (pin 8) through a 2.2 kΩ resistor and $CH2$ to the serial input (pin 1). Set the TTL clock frequency to 1 kHz. Set logic switch 1 to 0 and then to 1. Press **Autoset** on the oscilloscope. Rotate the *HORIZONTAL SCALE* knob until the time per division is 4 ms. Press either **Run/Stop** or **Single**. If the sequence stops cycling by becoming all ones (a state from which it cannot extricate itself) reset the display by setting logic switch one to zero and then back to one. Sketch or print the waveforms and demonstrate the properly functioning circuit to the laboratory instructor.

Laboratory Instructor Verification ______________________________

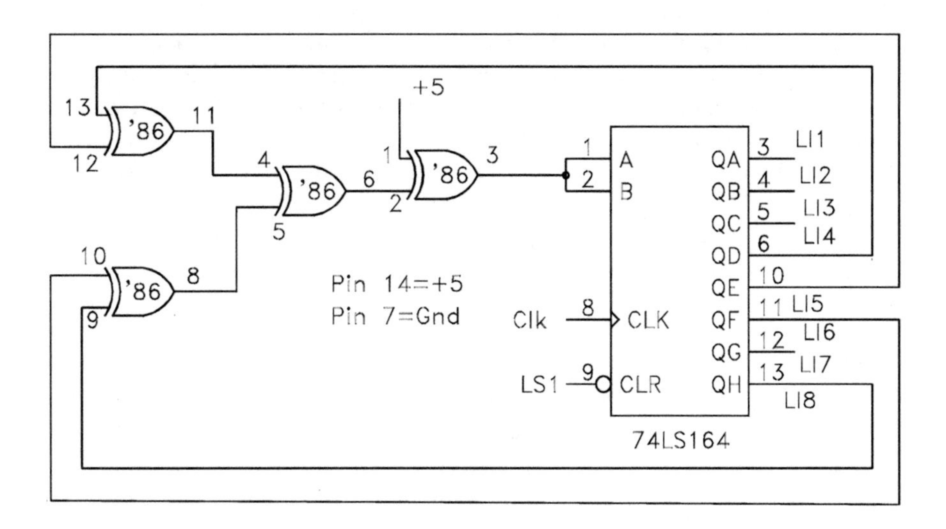

Figure 10.7: 8 stage pseudo-random sequence generator.

10.3.4 Eight Stage Pseudo-Random Sequence Generator

Assemble the circuit shown in Fig. 10.7 with the power off. Since all 8 stages are to be used, this will produce a pseudo-random sequence with a length $K = 2^N - 1 = 2^8 - 1 = 255$. This uses the same ICs that were used in the previous step. Leave the connection to the oscilloscope that were used in the previous step. Connect the TTL clock to pin 8 and set the frequency to 1 kHz.

Turn the power on. Switch logic switch one to 0 and then to one. Press **Autoset** on the oscilloscope. Change the time per division to 4 ms as before. Select either **Run/Stop** or **Single**. Note that it may not be possible to determine the period for the pseudo-random sequence (Why?). Print and describe the display.

____

____

Demonstrate the properly function circuit to the laboratory instructor.

Laboratory Instructor Verification ____

Turn the oscilloscope off. Set the TTL clock frequency to 10 Hz. Set logic switch one to 0 and then to 1 and measure with a wristwatch the cycle time of the pseudo-random sequence.

____

10.3.5 White Noise Source

Assemble the circuit shown in Fig. 10.8 with the power off. Begin by connecting pin 14 of each of the three ICs to +5 V and pin 7 of each of the three ICs to ground. Remove all connections to the logic state indicators. Two separate 74LS164s are required as well as one 74LS86. Connect the TTL clock to pin 8 of each of the 74LS164 shift registers. Connect logic switch 1 LS1 to pin 9 of each of the 74LS164 shift registers. A discrete 2N3904 NPN bipolar junction transistor is being used to obtain power gain for the speaker. (The lead placements for the 2N3904 are shown in the figure below the circuit diagram. Grasp the transistor so that the printed material on the flat face can be read. The emitter is then the lead on the left, the base the lead in the middle, and the collector the lead on the right.) The speaker is located on the right of the **CADET**. (Try not to get the two resistors backwards. It will probably destroy both the transistor and the shift register.)

This circuit should produce white noise. White noise would be a sound with all frequencies present in equal amplitudes just as white light consists of all colors present in equal amplitudes. The actual sound will not be white noise due to the poor frequency response of the speaker and the finite length of the pseudo-random sequence.

Turn on the power. Set the TTL clock frequency to 100 kHz. Describe the sound emanating from the speaker. (If no sound can be heard, reset the shift register by placing LS1 from the 1 to the 0 position and then back to the 1 position.)

____

____

____

Change the frequency of the clock to 10 kHz and describe the sound.

____

____

____

Change the frequency of the clock to 1 kHz and describe the sound.

____

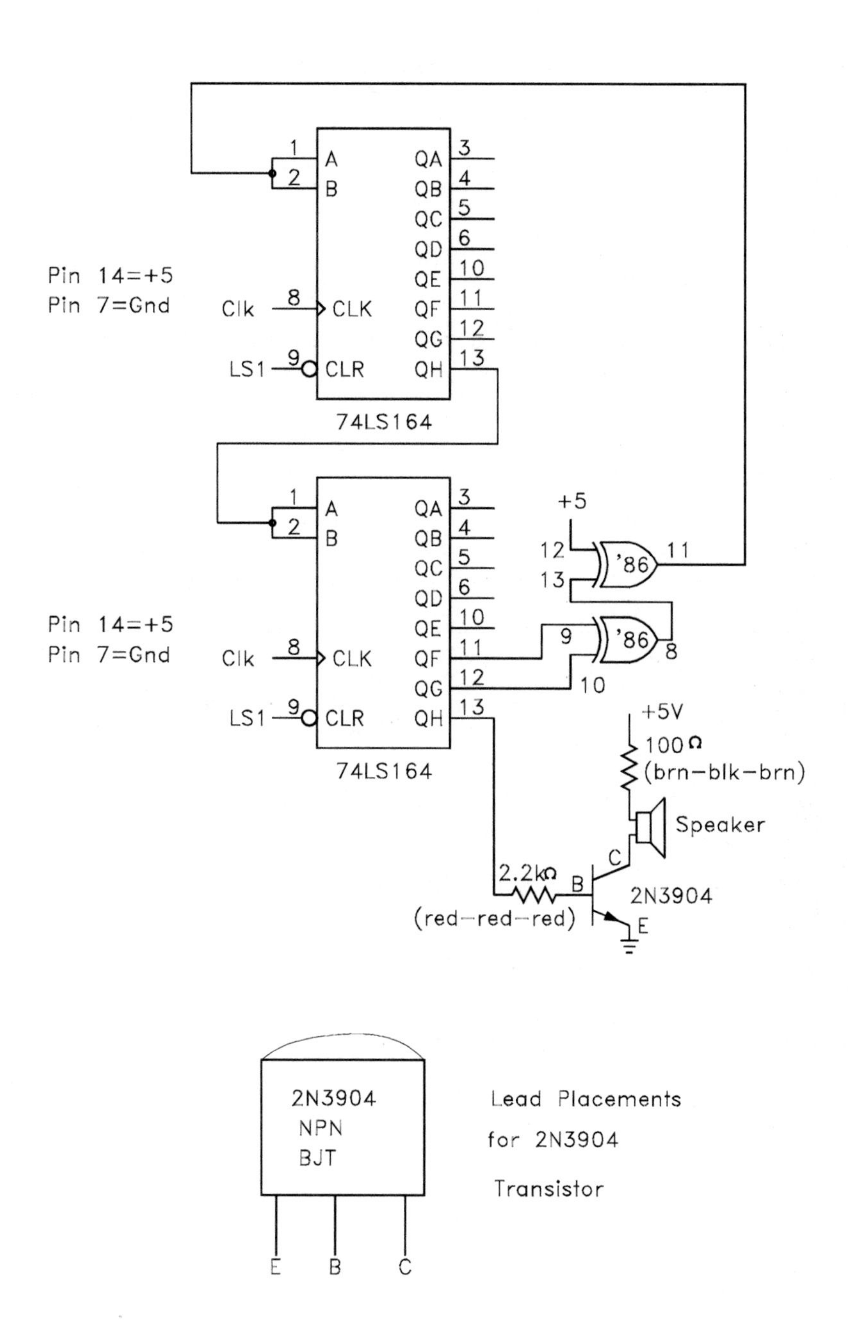

Figure 10.8: White noise source.

Demonstrate the properly function circuit to the laboratory instructor.

Laboratory Instructor Verification ______________________

Turn off the **CADET**.

10.4 Laboratory Report

1. What was the cycle time measured for the 4 stage and 8 stage pseudo-random sequences. Was this what would be expected? Why?

2. Why was it not possible to measure the period for the 8 stage pseudo-random sequence?

3. Why were the pseudo-random sequence generators that were examined not able to extricate themselves from the all ones state?

4. In step 10.3.5 the sound that should have emanated from the speaker should have sounded like random or white noise plus a periodic component. What should have been the frequency of the periodic component for the three clock frequencies 100 kHz, 10 kHz, and 1 kHz?

Turn in all data that was taken including all of the sketches. Turn in the answers to all questions as well as any supplementary questions that may have been posed by the laboratory instructor.

Title:____________________

Figure No.____________________

VOLTS/DIV

Ch1 __________
Ch2 __________
Ch3 __________
Ch4 __________

TIME/DIV
MAIN

____

DELAYED

____

DELAY

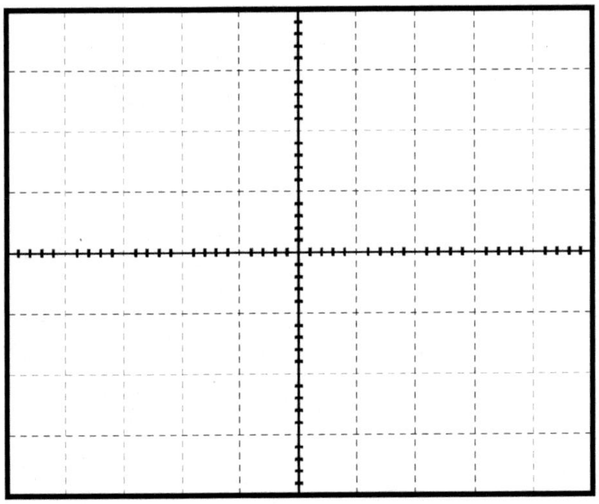

____

Coupling
AC ☐
DC ☐
Ch 2 Inv ☐
BW Limit ☐

Title:____________________

Figure No.____________________

VOLTS/DIV

Ch1 __________
Ch2 __________
Ch3 __________
Ch4 __________

TIME/DIV
MAIN

____

DELAYED

____

DELAY

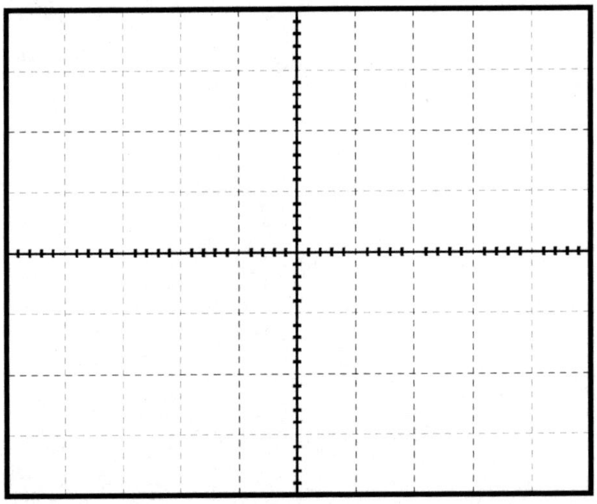

____

Coupling
AC ☐
DC ☐
Ch 2 Inv ☐
BW Limit ☐

Figure 10.9:

Chapter 11

Analog to Digital and Digital to Analog Conversion Systems

11.1 Object

The object of the experiment is to examine analog-to-digital and digital-to-analog conversion. The specific analog-to-digital converter that is the subject of this experiment is the tracking A/D converter.

11.2 Theory

11.2.1 Introduction

Oftentimes it is more convenient, economical, and efficient to store, process, or transmit signals in digital rather than analog form. The plethora of cheap and powerful microprocessors makes it possible to perform complex signal processing on digitized signals which would be almost impossible if they were in analog form. Additionally, when signals are in digital form they are less susceptible to the deleterious effects of additive noise which makes it desirable to transmit and store them in digital form.

The devices that transform a signal from analog to digital form are known as analog-to-digital converters (acronyms in use are A/D and ADC). When the analog signal is to be recovered from its digital representation a device that performs the reciprocal of analog-to-digital conversion is required which is simply known as the digital-to-analog converter (acronyms in use and D/A and DAC).

This experiment will examine a simple analog-to-digital conversion system which uses a digital-to-analog converter as an integral part of the conversion process. It will be constructed in component form so that each component of the system can be examined individually.

11.2.2 Analog-to-Digital Converter Fundamentals

An analog quantity may assume any of an infinite number of values between its maximum and minimum values. (For simplicity it will be assumed that the analog quantity being discussed is a voltage. Any analog quantity can be converted into a voltage by the choice of an appropriate transducer). The difference between the maximum and minimum value of the voltage is known as the range voltage, $V_r = V_{\max} - V_{\min}$. In order to perform an analog-to-digital conversion, this range must be subdivided into intervals and a digital code word or data word assigned to each interval. Obviously, if these intervals have equal length, the accuracy of the conversion will be a monotonically increasing function of the number of the intervals and consequently the number of code words used to represent the digitized signal.

The number of code words used to represent a signal is dependent on the length of the code word. If the natural binary number system is used, then the length of the code word is equal to the number of bits of information that it contains. For an N bit binary code, the number of available code words is 2^N. If the analog signal is to be divided into equal intervals for digitization, then the length of each interval corresponding to a code word is $V_{ref}/2^N$ where V_{ref} is known as the reference voltage of the A/D converter (The reference is the largest change in the input voltage that can be encoded. It will be assumed that the range voltage of the input signal and reference voltage of the A/D converter are the same). The length of each of the intervals determines the resolution of the A/D converter (the resolution is the smallest change in the input voltage that guarantees that the output code word will change) and this resolution is given by

$$\Delta V = \frac{V_r}{2^N} \tag{11.1}$$

where it is assumed that the range and reference voltage are the same (If it weren't a simple op amp circuit could be used to shift the level and make the variation of the input voltage to the A/D equal to the reference voltage.) The resolution represents the quantization error inherent in the conversion of the signal to digital form, i.e. it is impossible to reconstruct the analog signal from the digitized version with a precision greater than ΔV. The resolution is sometimes simply referred to as the LSB (least significant bit) in that it represents the voltage range corresponding to a change in the input code of one bit.

The dynamic range of an A/D converter is the ratio of the maximum to minimum voltage change (usually expressed in decibels) that the converter can encode and is given by $20\ \log_{10}(2^N)$ in decibels. This also provides a measure of the resolution of an A/D converter.

Another important parameter associated with A/D systems is the conversion time, sets a limit to the accuracy of the conversion process since the input signal must not change significantly while the conversion is in progress. If the signal changes by more than ΔV in a period of time equal to t_{conv}, then an error is guaranteed. This sets a limit to the ΔV in a period of time equal to t_{conv}, then an error is guaranteed. This sets a limit to the maximum frequency of a sinusoid that can be successfully digitized.

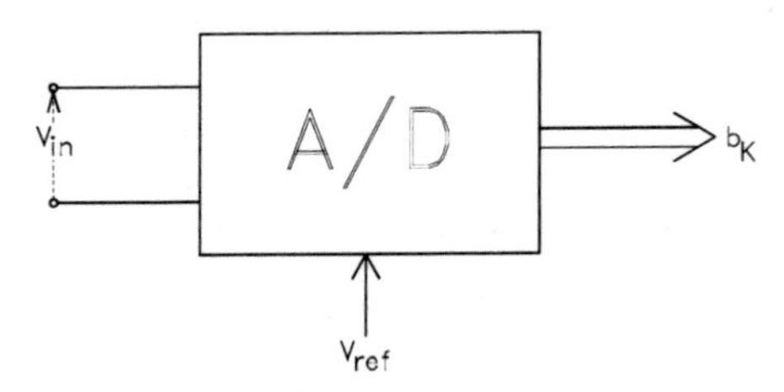

Figure 11.1: Analog to digital converter.

Fig. 11.1 illustrates a general A/D converter. As the input voltage varies from 0 to V_r the output code, b_k, varies from all zeros to all ones. Each code word b_k is of the form $b_k = (B_1 B_2 B_3\ ...\ B_N)$ where each B_i is either 0 or 1 and B_1 is the most significant bit (MSB) and B_N is the least significant bit (LSB). This type of binary code is called simple, straight, or natural binary. As V_{in} varies from 0 to V_{ref} the output code increases from 0 to $2^N - 1$ in binary increments of 1.

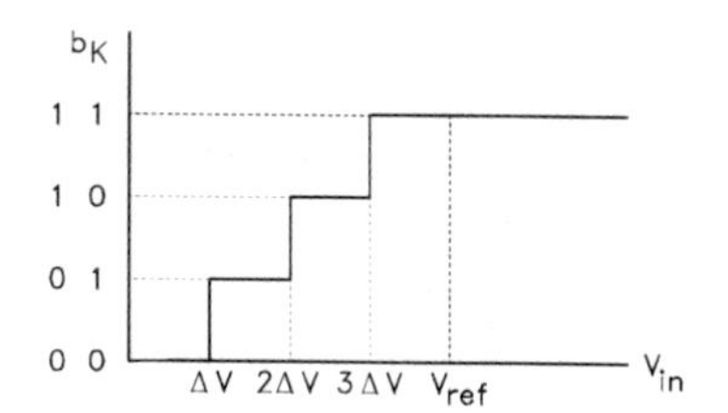

Figure 11.2: Transfer characteristic for 2 bit A/D

Fig. 11.2 illustrates the transfer characteristic of a 2 bit A/D converter. For simplicity it is assumed that $V_{\max} = V_{ref} = V_r$ and that $V_{\min} = 0$. The input analog voltage is converted into a digital code word as

V_{in}	b_k
$0 - \Delta V$	00
$\Delta V - 2\ \Delta V$	01
$2\ \Delta V - 3\ \Delta V$	10
$3\ \Delta V - V_r$	11

the resolution is $\Delta V = 0.25\ V_r$, and the dynamic range is 12 dB. A voltage less than zero would be encoded as all 0's and a voltage greater than V_r would be encoded as all 1's. Sophisticated A/D converters would also have indications of when the input voltage is above V_{ref} (over-range) and below 0 (under-range). Generalization to higher order (more bits) A/D converters is obvious and tedious. This type of A/D converter is known as an equal interval converter since the voltage intervals corresponding to each code word have equal lengths.

A variation of the above encoding process is also used in which the voltage levels are shifted by $\Delta V/2$. This would correspond to for the two bit A/D

V_{in}	b_k
$0 - .5\ \Delta V$	00
$0.5\ \Delta V - 1.5\ \Delta V$	01
$1.5\ \Delta V - 2.5\ \Delta V$	10
$2.5\ \Delta V - V_r$	11

which has the same resolution and dynamic range as the previous conversion scheme. Each interval for V_{in} has a length of ΔV except all 0's (length $\Delta V/2$) and all 1's (length 1.5 ΔV). This conversion scheme has the advantage of being slightly more linear than the previous one. This is the encoding scheme that is normally used in industrial applications. For A/D converters with a large number of bits the difference between these two coding schemes is slight.

The number of bits used in practical systems is rarely less than 8. That is because many of the cheaper microprocessors employ 8 bits and the cheapest commercially available A/D converters therefore also employ 8 bits. This results in a resolution of $\Delta V = V_r/256$ and a dynamic range of 48 dB.

Compact disc recording systems employ 16 bit A/D converters to encode music. The results in a dynamic range of 96 dB which for sound corresponds to the difference between the largest and smallest sound that can be reproduced.

11.2.3 Digital-to-Analog Converters

Digital-to-analog converters (D/A) are required when it is desired to convert a digital signal to analog form. They are often integral parts of an A/D converter so it is logical to consider them first.

Code words are applied to the input of the D/A and an analog voltage is produced at the output. An N bit D/A converter produces one of 2^N analog output voltages for each of the possible 2^N input binary code words. Many of the terms, such as resolution and reference voltage, used to describe A/D converters are equally applicable to D/A converters.

The transfer characteristic of a D/A converter is normally the complement of the transfer function of the corresponding A/D converter. The analog output voltage is a monotonically increasing function of the input binary code word. Normally an input code word of all 0's would produce an output voltage of zero and an input of all 1's would produce an output voltage of $V_{ref}(1 - 2^{-N})$ which is the reference voltage minus the resolution. (If a bipolar output were desired, it could be obtained by using an op-amp as a level shifter.)

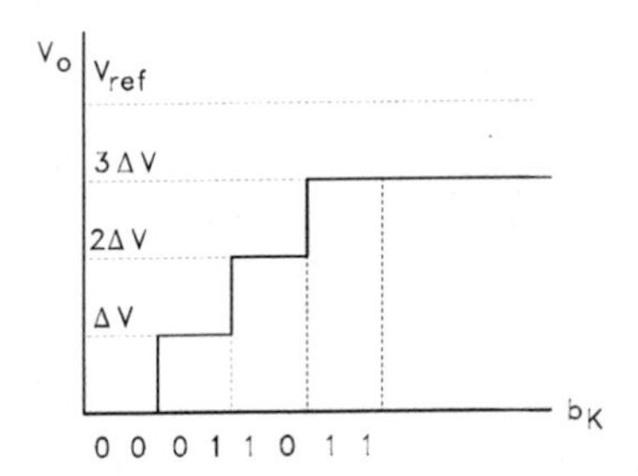

Figure 11.3: Transfer characteristic for 2 bit digital to analog converter.

Shown in Fig. 11.3 is the transfer characteristic of a 4 bit D/A. The output voltage is given as a function of the input code by

b_k	V_o
00	0
01	ΔV
10	$2\ \Delta V$
11	$3\ \Delta V$

where ΔV is the resolution of the D/A converter given by $\Delta V = V_{ref}/2^N = V_{ref}/4$. It is important to note that an input code of all 1's does not produce the reference voltage; it produces the reference voltage minus the resolution voltage. For D/A converters employing a large number of bits, the input code word of all 1's would have an output voltage that would be approximately the reference voltage.

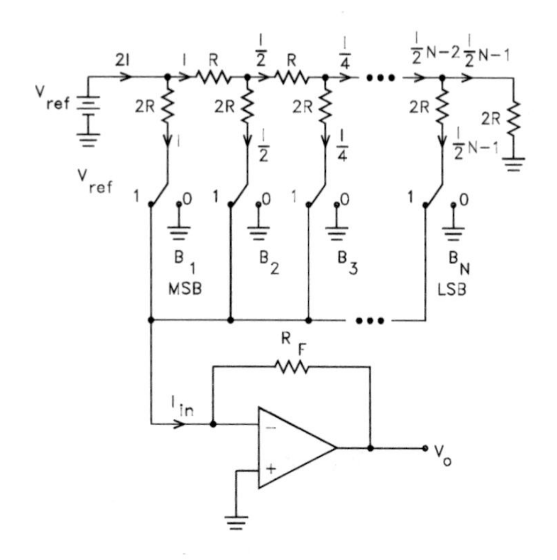

Figure 11.4: An N bit $R-2R$ ladder D/A converter.

Digital-to-analog converters are normally implemented using a circuit known as the $R-2R$ ladder. Such a circuit is shown in Fig. 11.4 for an N bit D/A converter. Only two resistor values are needed to implement this circuit and they can be precisely made. The input binary code word $b_k = (B_1B_2B_3...B_N)$ determines the setting of the N switches which determines whether the current flowing through each $2R$ resistor flows to ground or the lead connected to the inverting input of the op-amp. This circuit has the property that the resistance seen looking to the right from just to the left of the top N nodes is always R and the current splits in half at each node. (The currents in the $2R$ branches are not functions of the switch positions since the switch is either set to ground [0 position] or virtual ground [1 position]) The current I_{in} is then given by

$$I_{in} = I\sum_{i=1}^{N}\frac{B_i}{2^{i-1}} \tag{11.2}$$

where the current I is given by $I = V_{ref}/(2R)$. Since the current I_{in} flows through the feedback resistor of the op-amp (ideal op-amp assumption invoked), the output voltage of the D/A converter is

$$V_o = V_{ref}\frac{R_F}{2R}\sum_{i=1}^{N}\frac{B_i}{2^{i-1}} \tag{11.3}$$

where the input code word b_k determines the $B_i's$ which are either 0 or 1. If $R_f = R$, then the output voltage varies from 0 volts when the input code is all 0's to $V_{ref}(1-2^{-N})$ when the input code word is all 1's. This circuit would normally be contained in a single monolithic integrated circuit with inputs for the N bit code word and the reference voltage; the switches are electronic rather than mechanical. The accuracy of the D/A converters (or A/D) can never be more accurate than the precision of the reference voltage. For this reason, a special circuit known as a voltage reference would normally be used with a D/A converter.

Assuming $R_F = R$ the expression for the output voltage can be written as

$$V_o = V_{ref}\sum_{i=1}^{N}\frac{B_i}{2^i} = \frac{V_{ref}}{2^N}\sum_{i=1}^{N}2^{N-i}B_i \tag{11.4}$$

The output voltage when the input is all 0s is zero. The output voltage when the input is all 1s is known as the full scale voltage, V_{FS}. The full scale voltage is given by

$$V_{FS} = \frac{V_{ref}}{2^N} \sum_{i=1}^{N} 2^{N-i} = V_{ref} \frac{2^N - 1}{2^N} = V_{ref} \left(1 - 2^{-N}\right) \tag{11.5}$$

Thus the output voltage of the DAC goes from 0 V to V_{FS} as the input binary code goes from all 0s to all 1s. If a plot is made of the output voltage as a function of the input code it should be a straight line. This linearity is one of the most important properties of a DAC.

11.2.4 Analog-to-Digital Converters (A/D)

Numerous schemes are used to accomplish analog-to-digital conversion. Only a few of the most commonly used will be examined: dual slope integration, successive approximation, flash, and tracking. Some others are voltage-to-frequency converters, staircase ramp or single slope, charge balancing or redistribution, switched capacitor, delta-sigma, and synchro or resolver.

Dual Slope A/D Converter

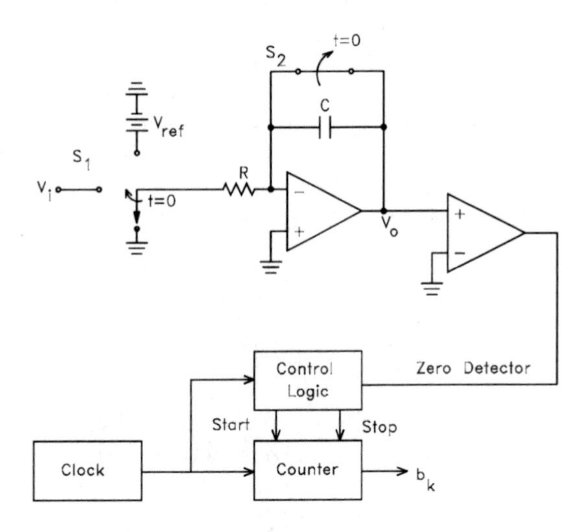

Figure 11.5: Dual slope A/D converter.

The dual slope A/D converter is shown in Fig. 11.5 in block diagram form. Its fundamental components are an integrator, electronically controlled switches, a comparator, a counter, a clock, a reference voltage, and some control logic.

Prior to the beginning of a conversion at $t = 0$, switch S_1 is set to the ground position and switch S_2 is closed which shorts the capacitor in the integrator. At $t = 0$ a conversion begins and switch S_2 is opened and switch S_1 is set to the center position which makes the input to the integrator V_{in}, the analog voltage to be digitized. Switch S_1 is held in the center position for an amount of time equal to T_i which is a constant predetermined time interval. Prior to time $t = 0$ the counter is set to 0 and when the switches are thrown at $t = 0$ the counter begins to count the clock pulses. The counter is designed to reset to zero at the end of the time interval T_i.

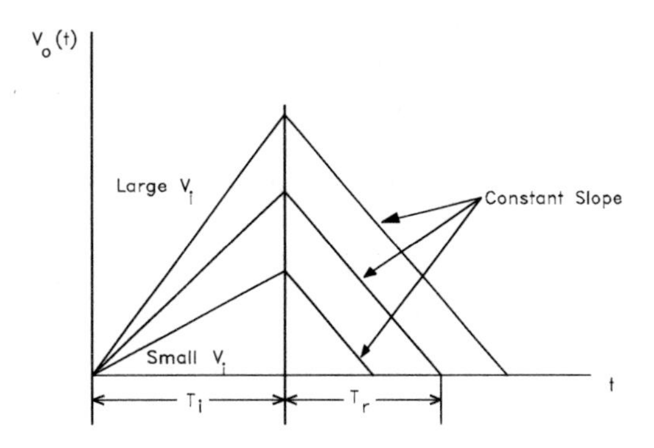

Figure 11.6: Integrator voltage for dual slope A/D converter.

The value of the output of the integrator voltage at time $t = T_i$ is $V_{in}T_i/RC$ which is linearly proportional to the input voltage, V_i. The output voltage of the integrator is shown in Fig. 11.6. The output voltage as a function of time is plotted for three different values of V_{in}. The output voltage is a straight line with a slope proportional to V_{in}.

At time $t = T_i$, after the counter has been set to zero, the switch S_1 is set to connect the voltage $-V_{ref}$ to the input of the integrator which has the voltage $V_{in}T_i/RC$ stored on it. The integrator voltage then decreases as a linear function of time with slope $-V_{ref}/RC$. A comparator is used to determine when the output voltage of the integrator crosses zero. When it is zero the input to the counter from the clock is disabled. The state of the counter is then the digitized value, b_k, of V_{in} since its count is then determined by T_r, the amount of time required for the integrator voltage to go from its value at $t = T_i$ to zero. The time interval T_r is then given by

$$T_r = \frac{V_{in}}{V_{ref}} T_i \tag{11.6}$$

which accomplished the analog-to-digital conversion since the times T_i and T_r are digital quantities (the number of pulses counted by the counter times the period of the clock) while the voltages are analog quantities (the uncertainty in this measurement of T_r is the period of the clock which means that the resolution of this converter is a function of the clock period). If N_i is the count at which the counter resets to 0 and N_r is the count of the counter when counter is disabled when the comparator detects zero crossing for the integrator output voltage, then the above equation becomes

$$N_r = \frac{V_{in}}{V_{ref}} N_i \tag{11.7}$$

where the brackets indicate the first integer less than the quantity inside the brackets.

The most common use of this type of A/D is in digital voltmeters because of the noise rejection characteristics of the integrator. One of the more troublesome noises that can appear at the input of a digital voltmeter is 60 Hz AC power line ripple and its harmonics. By selecting the time interval T_i to be the period of a 60 Hz sinusoidal, 16.67 ms, any 60 Hz noise and its harmonics will be integrated out at the end of the time interval T_i. If the counter were designed to count 3,000 pulses and then reset to zero at $t = T_i$, then the clock frequency would be 179.96 kHz.

If the reference voltage were picked to be 30 mV, the count N_r would be the digitized value of the input voltage. For example if the input voltage were 10 mV, then the count N_r would be 1,000 which is the input voltage except for the decimal place. If a decade (four BCD counters) rather than binary counter were used, then this would be a 3 and 1/2 digit digital voltmeter since the last three decimal digits could range from 0 to 9 but the most significant decimal digit could only range from 0 to 3.

Ranges larger than 30 mV could be accommodated by placing an input attenuator at the input to the A/D converter. Since the largest value T_r can have is T_i this means that if a zero crossing is not obtained after $2T_i$, then the input voltage V_{in} is larger than V_{ref} and an over-range indication could be given by the control logic.

The conversion time for this A/D converter ranges from T_i to 2 T_i. If T_i is picked as 16.67 ms, this is too long for some applications. This type of A/D converter is used in digital voltmeters where speed is not a high priority but accuracy and noise rejection are, i.e. measurement of DC voltages. The accuracy of this A/D converter is determined by the precision of the clock frequency and the reference voltage.

Successive Approximation A/D Converter

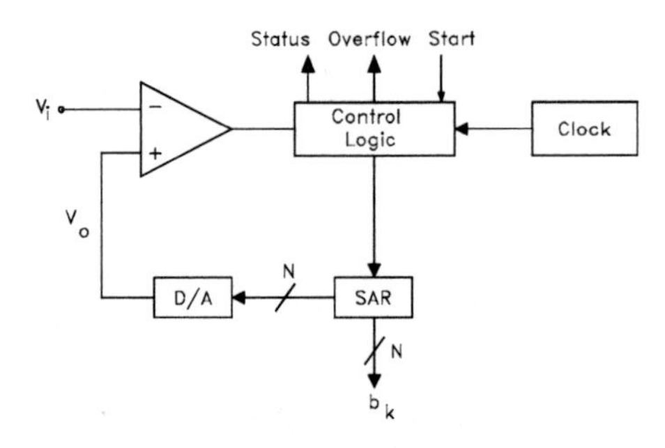

Figure 11.7: Successive approximation A/D converter.

A block diagram of a successive approximation A/D converter is shown in Fig. 11.7. The components of this converter are a successive approximation register (SAR), a clock, a comparator, a D/A converter, and some control logic. For an N bit A/D converter, an N bit D/A is required and an N stage register. This is a feedback system in which the output of the D/A converter is compared with the voltage V_{in} to be digitized until the difference between the two is sufficiently small.

The conversion begins a start input to the control logic. When this happens the first clock pulse sets the state of the SAR to the code word $(100\cdots00)$ for which the most significant digit is one and all the rest are zero. This makes the input to the D/A the code word which produces an output voltage $V_o = V_{ref}/2$. The output of the comparator then determines whether V_{in} is greater than or less than one-half of V_{ref}. If V_{in} is greater than $V_{ref}/2$ then $B_1(MSB)$ is set to 1, and if the converse is true then B_1 is set to 0.

On the next clock pulse, the input to the D/A converter is set to $(B_1 1\ 0\ 0\cdots00)$, and the voltage V_o is compared to V_{in}. If V_{in} is greater than V_o then B_2 is set to 1 and if the converse is true B_2 is set to 0. This process continues until the value of all N bits has been determined. The status output line then indicates that the A/D has been completed. If at the end of the determination of the LSB, B_N, the voltage V_{in} is still larger than V_o an overflow indication is given.

This type of A/D converter is analogous to weighing an unknown weight with a balance pan type of scale. The unknown weight would be place in one pan and then the heaviest standard weight would be placed in the other pan. If the unknown weight were larger than the heaviest standard weight, then the heaviest standard weight would be left in the pan ($MSB = 1$) and if the converse were true then the heaviest standard weight would be removed ($MSB = 0$). The next heaviest weight would be added to the pan and the process repeated until the lightest weight was reached. The weight of the lightest weight would represent the resolution of this analog-to-digital conversion.

The conversion time of this A/D is $N\ T$ where T is the period of the clock. If a high speed clock is used, this A/D can have a relatively short conversion time. The actual value of the conversion time in an actual successive approximation A/D would be slight larger than $N\ T$ due to the fact that the comparator and D/A converter have non-negligible settling times. Also in a practical converter, a type D latch flip-flop would be used to deglitch the output of the D/A and, since the voltage V_{in} must remain constant during the conversion cycle or an error will occur, a sample-and-hold circuit would precede the input to the A/D. This is a widely used scheme for A/D conversion and is usually placed on a single monolithic integrated circuit.

Flash A/D Converter

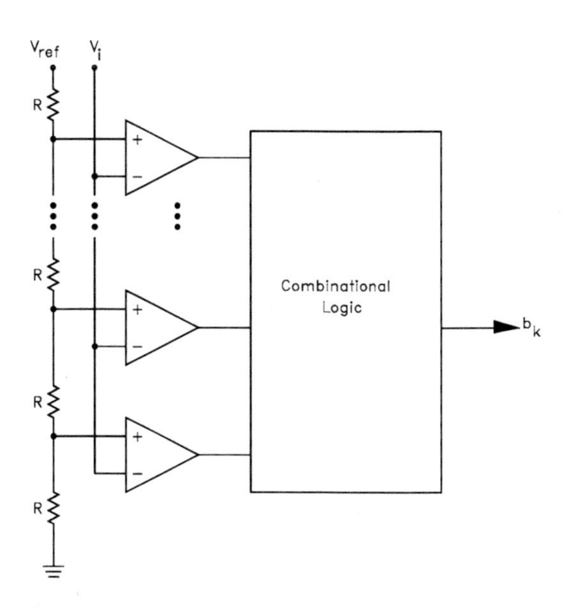

Figure 11.8: Flash A/D converter.

A block diagram of a flash or parallel A/D converter is shown in Fig. 11.8. The component parts required for an N bit A/D flash converter is $2^N - 1$ comparators, 2^N equal resistors, and some combinational logic.

This comparator simply uses the 2^N resistors to form a ladder voltage divider which divides the reference voltage into 2^N equal intervals and the $2^N - 1$ comparators to determine in which of these 2^N voltage intervals the input voltage V_{in} lies. The combinational logic then translates the information provided by the output of the comparators into the code word b_k.

This converter does not inherently require a clock so the conversion time is essentially set by the settling time of the comparators and propagation time of the combinational logic. This means that this is the fastest of the A/D converters and is used when minimum conversion time is required. In practice, a sample-and-hold circuit would be used at the input and D type latch flip-flops at the output. A two cycle clock would be used to sample the input on the first cycle and open the latches on the second half cycle after everything has settled.

Tracking A/D Converter

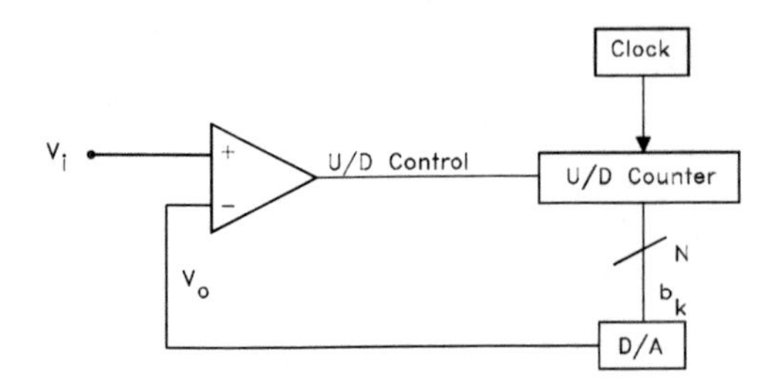

Figure 11.9: Tracking A/D converter.

A block diagram of a tracking A/D converter is shown in Fig. 11.9. The components of this converter are a comparator, an up-down counter, a clock, and a D/A converter. This is another feedback A/D converter which uses a comparator to examine the input voltage, V_{in}, and the output voltage V_o of a D/A converter. If an N bit D/A converter is used, then, in general, this results in an $(N-1)$ bit A/D converter. (The LSB of the D/A alternates between 1 and 0 on alternate clock cycles and the output of the A/D converter is the first $N-1$ most significant bits of the D/A converter.) If the state of the counter is sampled when the counter is in its count down state, then this can be used as an N bit A/D converter.

The counter in this converter is always counting the pulses produced by the clock. When the input voltage is larger than the D/A output voltage the comparator cause the counter to count up and when the converse is true the counter counts down. The analog to digital conversion is performed by the comparator; the comparator output indicates whether the voltage V_{in} is larger or smaller than V_o. When the magnitude of the difference between V_{in} and V_o is less than $V_{ref}/2^{N-1}$ where V_{ref} is the reference voltage of the D/A converter, then the conversion is complete.

The LSB of the D/A oscillates between 0 and 1 on alternate clock cycles. If V_{in} is larger than V_o but this difference is less than the resolution of the A/D converter ($V_{ref}/2^{N-1}$), then the counter counts up on the next clock cycle which makes V_{in} less than V_o on the next clock cycle. This process of counting up and down by one LSB of the D/A converter repeats ad infinitum.

The conversion time of this A/D converter depends on the difference between V_{in} and V_o. The maximum value of the conversion time is $T\ 2^N$ when it is in its acquisition mode and T when it is in its tracking mode where T is the period of the clock. By picking the clock frequency to be sufficiently high, the conversion time can be made reasonably small.

This is the A/D converter that will be the subject of this experiment. It will be constructed in discrete form so that the performance of each component can be examined. This would be a poor choice for a practical A/D converter in that it is susceptible to noise pickup and some of the sophisticated circuitry that should be used, such as a sample and hold circuit, will not be employed.

11.2.5 D/A Converter

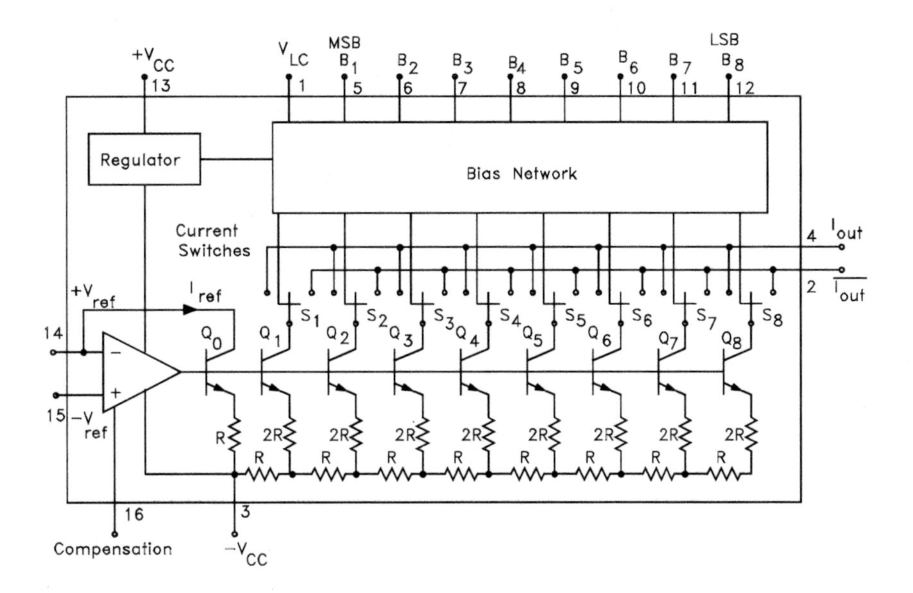

Figure 11.10: DAC-08 D/A converter.

The D/A converter that will be used in this experiment will be the DAC-08. It is an 8 bit TTL device that is intended to be directly interfaced with an 8 bit microprocessor. The MSB is pin 5 and the LSB is pin 12 as shown in the diagram in Fig. 11.10.

The DAC-08 is an $R-2R$ ladder D/A. A voltage is applied at pin 14 which produces the reference current flowing into pin 14. This current I_{ref} is the current $2I$ discussed previously; it supplies the input current to the $R-2R$ ladder. The output current I_{out} flowing into pin 4 is given by

$$I_{out} = I_{ref} \sum_{i=1}^{8} \frac{B_i}{2^i} \tag{11.8}$$

where B_1 is the MSB and B_8 is the LSB. The optimum value of I_{ref} is 2 ma. The accuracy of the current I_{ref} is the fundamental factor determining the accuracy of the A/D converter to be constructed using this device.

The output voltage is given by

$$V_o = \frac{V_{ref}}{2^8} \sum_{i=1}^{8} 2^{8-i} B_i \tag{11.9}$$

and the full scale voltage (the output voltage when the input is the binary code word consisting of all 1s) is

$$V_{FS} = \frac{255}{256} V_{ref} \tag{11.10}$$

Thus the output goes from 0 V to V_{FS} in 256 steps as the input code word goes from all 0s to all 1s.

11.3 Procedure

11.3.1 Buses

Establish buses or rails for the breadboard. Four rails are required: +12 V, −12 V, 5 V, and ground. Both op amps and TTL digital ICs will be used which require these four voltages. Use the long vertical buses that are parallel to the IC trough. Be sure to connect jumpers at the half way point so that the entire breadboard can be used. Turn on the **CADET** and the **Agilent 34401A** DMM. Adjust the variable positive and negative power supplies for +12 V and −12 V respectively. Turn off the **CADET**.

11.3.2 DAC

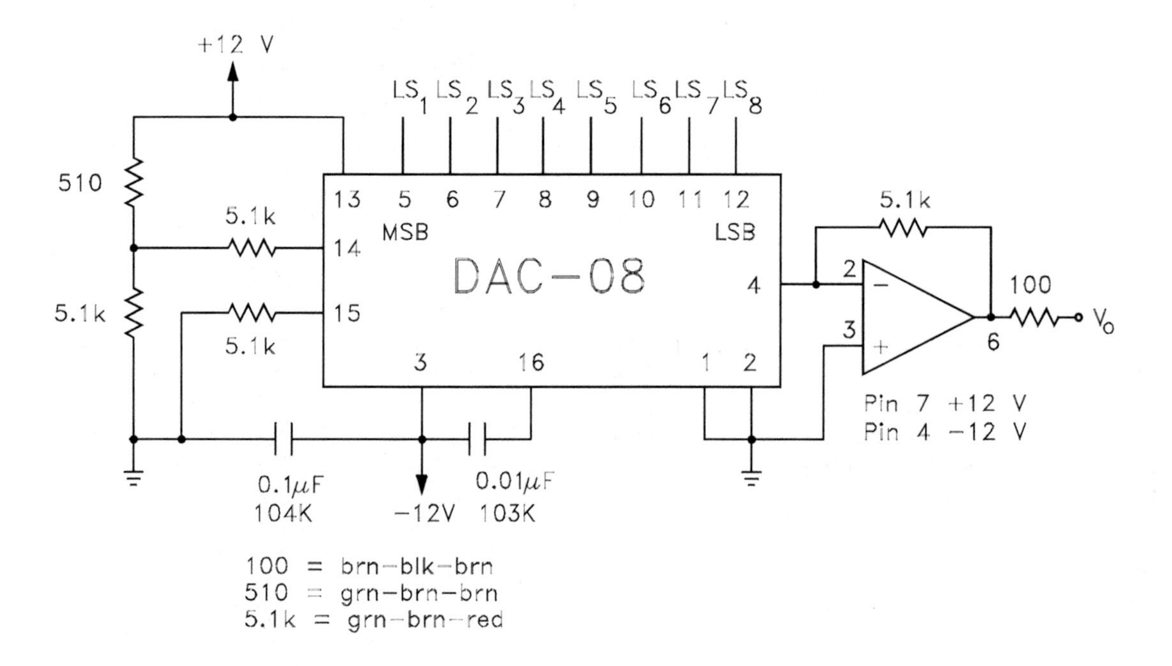

Figure 11.11: Digital to analog converter.

Assemble the circuit shown in Fig. 11.11 with the power on the **CADET** turned off. Use either a 741 or TL071 for the op amp. Be certain to connect pin 4 of the op amp to the −12 V rail and pin 7 to the +12 V rail. Logic Switch 1 is connected to the MSB (most significant bit) on the DAC-08 which is pin 5. The other pin connections are shown. The Logic Switches will be used to supply an input 8 bit code word to the DAC-08. Make sure that the switch to the left of the breadboard section for the logic switches is set to +5 V. Connect V_o to BNC 2 on the **CADET** and use a BNC to banana plug lead to connect BNC 2 to the input of the **Agilent 34401A** DMM configured as a DC voltmeter. Connect the red banana plug to the HI input and the black banana plug to the LO input.

Turn on the **CADET.** Set all the logic switches to 0 and measure the DC voltage at the output; it should be close to zero. Set all the logic switches to 1 and measure the output; it should be close to 10 V. If this is not the case there is a problem with the circuit.

Measure and record the output voltage (in the third column of the following table) for the following input binary codes:

Binary Code	N	Measured Voltage	Theoretical Voltage
00000000			
00000001			
01010101			
01000000			
01000100			
01001100			
10000000			
10000010			
10001100			
11000000			
11001100			
11100000			
11111111			

Laboratory Instructor Verification ______________________

Compute the reference voltage, V_{ref} as

$$V_{ref} = \frac{256}{255} V_{FS} \tag{11.11}$$

where V_{FS} is the full scale voltage which is the output dc voltage when the input binary code is all 1s. Use the value recorded in the above table for V_{FS} (it should be approximately 10 V).

Compute N for the binary codes used and record it in the second column of the above table. N is the decimal equivalent of the binary codes and is given by

$$N = B_1 2^7 + B_2 2^6 + B_3 2^5 + B_4 2^4 + B_5 2^3 + B_6 2^2 + B_7 2 + B_8 \tag{11.12}$$

where the input binary word is $b_N = (B_1 B_2 B_3 B_4 B_5 B_6 B_7 B_8)$ with B_1 the most significant bit and B_8 the least significant bit. The calculation may also be performed with a calculator which converts numbers from the binary to the decimal number systems.

Compute the theoretical voltage as

$$V_o = V_{ref} \frac{N}{256} \tag{11.13}$$

Plot the experimentally measured and theoretically calculated output dc voltages as a function of N on linear graph paper.

11.3.3 Stair Step

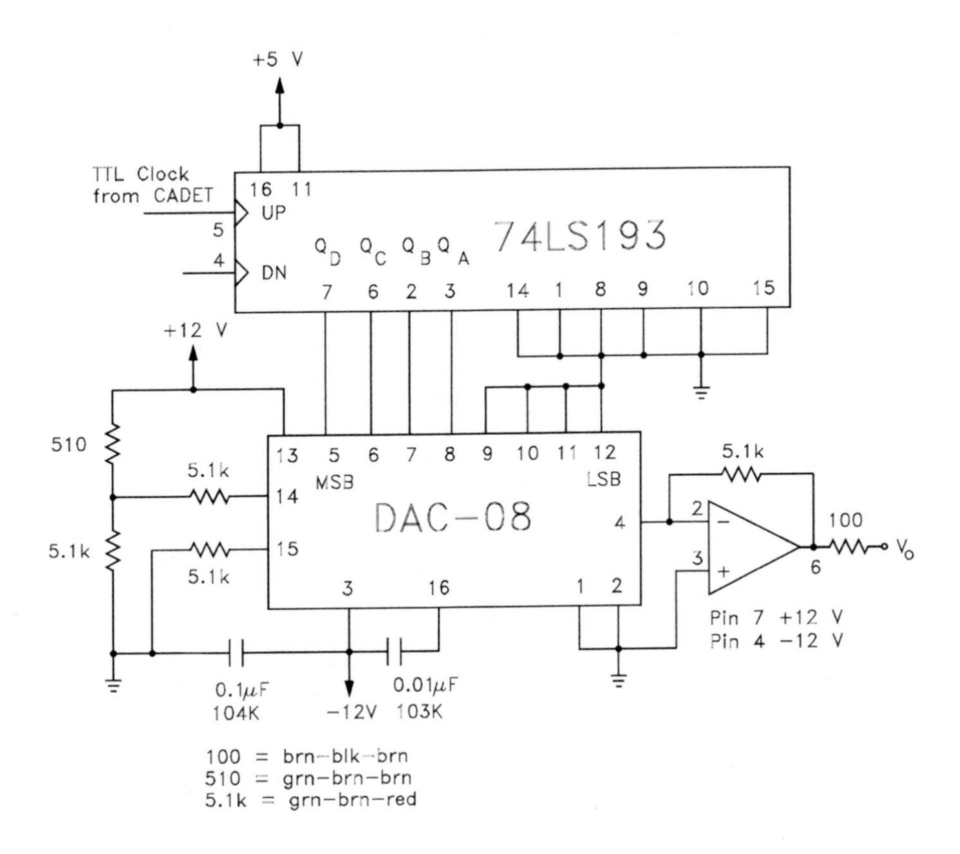

Figure 11.12: Stair step.

Assemble the circuit shown in Fig. 11.12 with the power on the **CADET** turned off. Be certain that pin 16 of the 74LS193 counter is connected to the +5 V supply rail and not the $V^+ = 12\,\mathrm{V}$ or it will be incinerated. Connect the TTL output of the **CADET** to pin 5 of the 74LS193. Also connect the TTL output of the **CADET** to BNC 1 through a 2.2 kΩ resistor (red-red-red).

Turn on the **Tektronix DPO3012** oscilloscope and wait for it to boot. Turn on the **CADET**. Connect BNC1 on the **CADET** to $CH1$ on the oscilloscope and BNC 2 to $CH2$. Turn $CH1$ off. Press **Autoset**. Use the $HORIZONTAL$ and $VERTICAL$ scale to obtain a suitable display that illustrates the stair step. Press either **Run/Stop** or **Single**. Use the Volt/Div setting to measure the height of each step. Print the display. Switch the TTL clock input to pin 4 of the 74LS193 and repeat.

How many steps are there on each set of stairs?

____

Laboratory Instructor Verification ______________________

11.3.4 Amplitude Hopping

Assemble the circuit shown in Fig. 11.13 with the power on the **CADET** turned off. Connect the TTL clock in series with a 2.2 kΩ (red-red-red) resistor to BNC 1 on the **CADET** and BNC 1 to $CH1$ on the oscilloscope. Connect the TTL clock from the **CADET** to pin 8 on the 74LS164 and logic switch one to pin 9 on the 74LS164. Set the frequency of the TTL clock to 1 kHz. Turn the power on. Switch logic switch one to 0 and then to 1. Turn $CH1$ off and $CH2$ on. Press **Autoset** on the oscilloscope. Use the $HORIZONTAL\ SCALE$ to change the time/div to 4 ms. Press **Single**. Press **Single** several times and note the change in the display.

Since the input to the DAC is a pseudo random sequence the output signal jumps or hops to different amplitude levels. Use the voltage per division to measure the height of the smallest hop. Print the display.

Laboratory Instructor Verification ______________________________

Turn the **CADET** off. Connect a 12 kΩ (brn-red-org) resistor from the +12 V power supply rail to pin 2 of the op amp (in addition to the components that are already there). Turn on the **CADET**. Press **Default Setup**. Turn $CH1$ off and $CH2$ on. Press **Autoset** on the oscilloscope. Use the $HORIZONTAL\ SCALE$ to change the time/div to 4 ms. Press **Single**. Press **Single** several times and note the change in the display. Press the spyglass icon and rotate the Pan Zoom knob and note effect on the display. Turn the spyglass off.

____

11.3.5 Spectral Analysis

The oscilloscope will now be configured to display the frequency spectrum of the waveform connected to output of the DAC. Press **Autoset** and then change the time per division to 40 ms with the $HORIZONTAL\ SCALE$ knob. Turn on the $MATH$ function, select FFT, and be sure the FFT source is set to $CH2$. Use the vertical position and scale controls to place $CH2$ in the upper half of the display and the FFT in the lower half so that they don't overlap. Turn on the cursors and position them so that the nulls in the spectra may be determined. Print the display.

Connect the output of the DAC, v_o to the top of the speaker on the **CADET**. Connect the bottom of the speaker to the ground on the **CADET**. Listen to and describe the sound emanating from the speaker.

____

Laboratory Instructor Verification ______________________________

11.4 Laboratory Report

From the plot of the output voltage as a function of input code, how linear is the DAC-08?

____

What were the height (in Volts) and width (in seconds) of the steps in the stairstep?

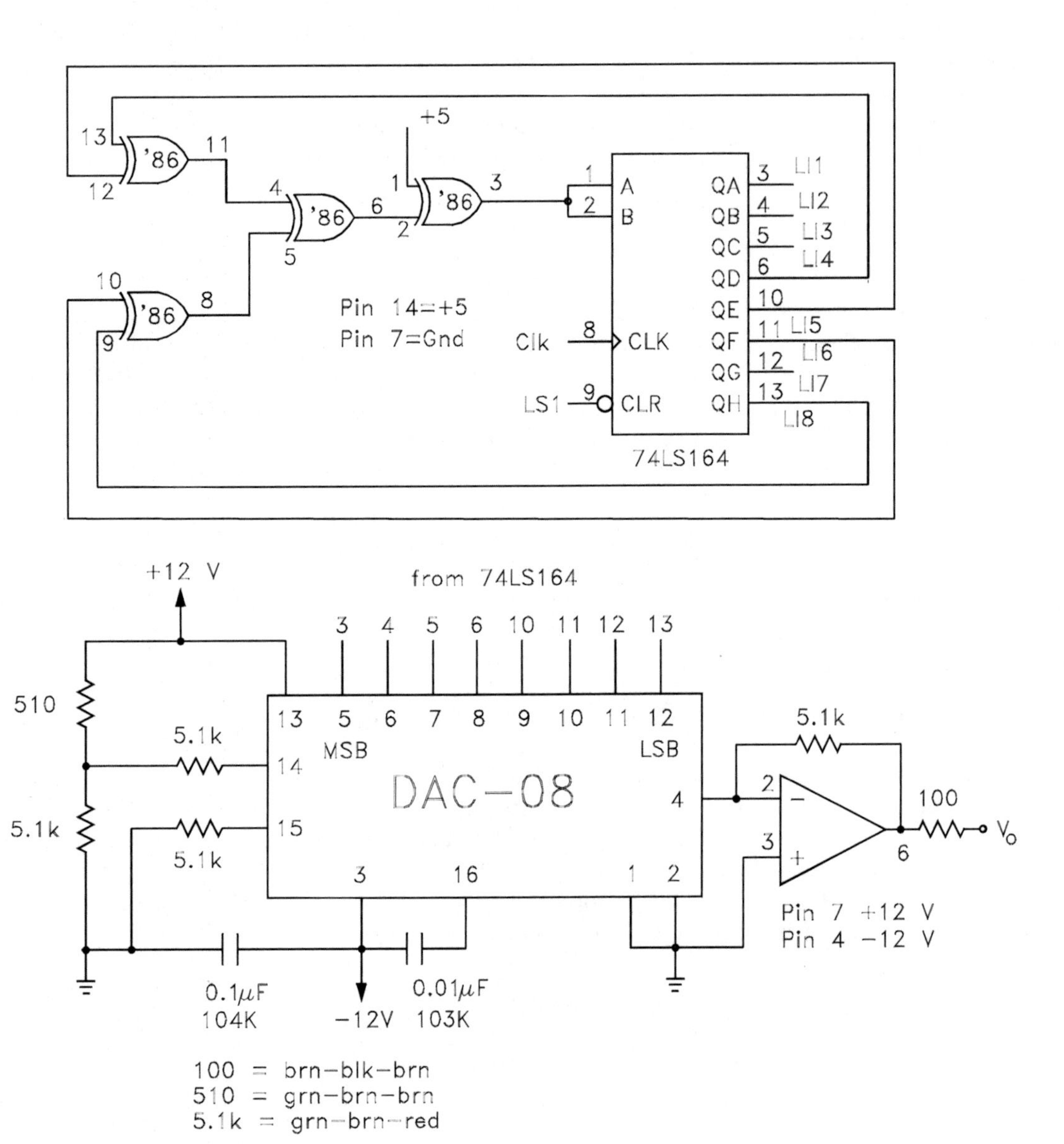

Figure 11.13: Amplitude hopper.

Examining the spectra of the amplitude hopper, why would a variation of such a circuit be found in cell phones? (With somewhat smaller and more elaborate electronics.)

Turn in all sketches, plots, and data taken. Answer all questions that were posed and any additional questions posed by the laboratory instructor.

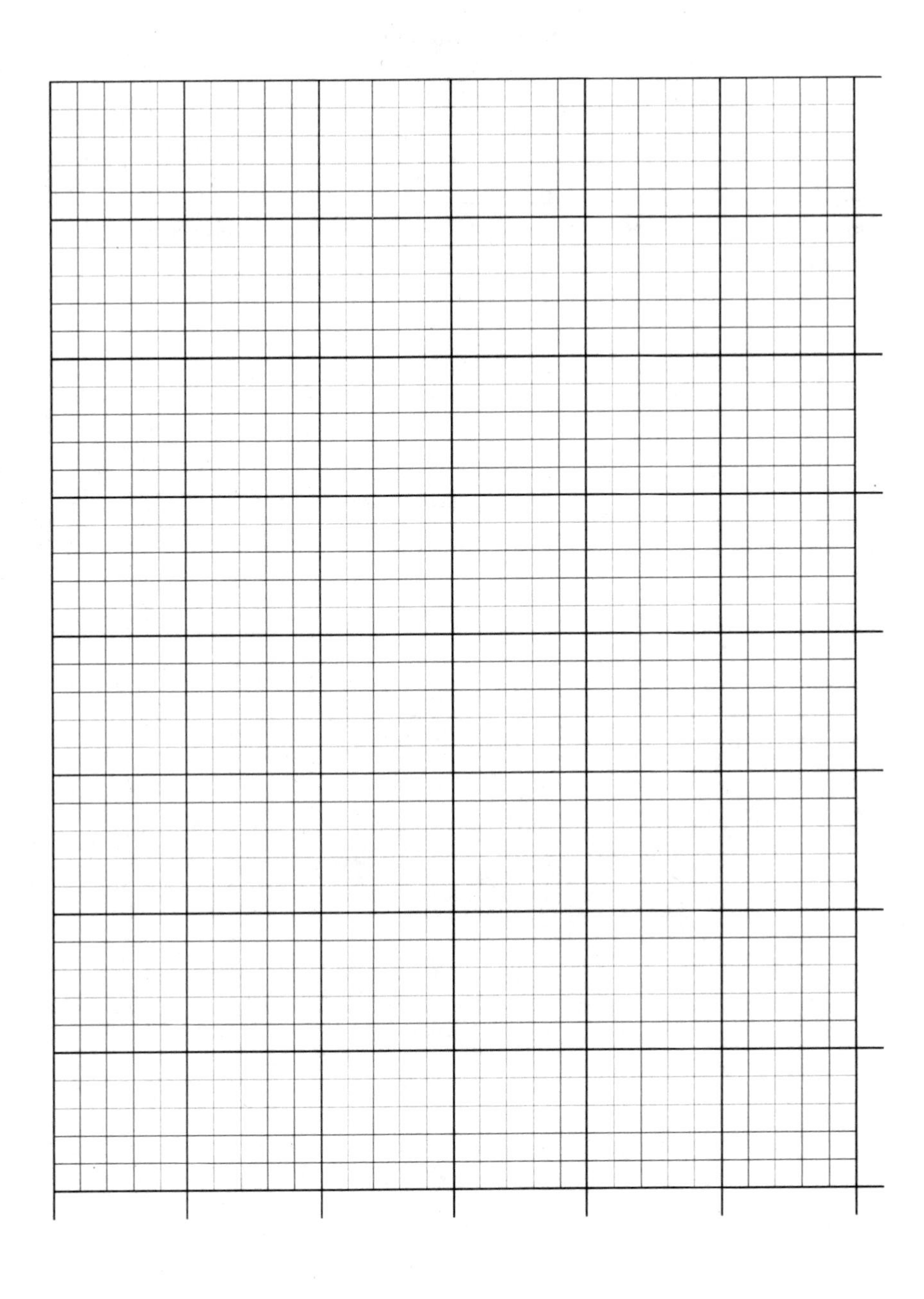